ENCYCLOPAEDIA OF EMINENT THINKERS

VOLUME-XVI: **SUBHAS CHANDRA BOSE**

Volumes by the same author

The Political Thought of ***Dadabhai Naoroji***
The Political Thought of ***Tej Bahadur Sapru***
The Political Thought of ***C. Rajagopalachari***
The Political Thought of ***Subhas Chandra Bose***
The Political Thought of ***Bhagat Singh***
The Political Thought of ***Annie Besant***
The Political Thought of ***C.R. Das***
The Political Thought of ***Lalalajpat Rai***
The Political Thought of ***Sir Pherozeshah Mehta***
The Political Thought of ***Mahadev Govind Ranade***

About the Author

Dr. (Miss) Ashu Pasricha is a faculty in Gandhian studies at Panjab University, Chandigarh. She has many books to her credit including *Gandhian Approach to Integrated Rural Development; Public Administration: Discipline and Dimensions; Peace Studies: Discipline and Dimensions; WTO, Self-reliance and Globalisation; and Gandhi's Concept of State and Power.* Also, she has published a number of research papers in various Indian and foreign journals of repute.

ENCYCLOPAEDIA OF EMINENT THINKERS

VOLUME SIXTEEN

POLITICAL THOUGHT OF SUBHAS CHANDRA BOSE

Dr. Ashu Pasricha
Professor
Department of Gandhian Studies
Panjab University, Chandigarh

CONCEPT PUBLISHING COMPANY, NEW DELHI-110059

Cataloging in Publication Data—DK
Courtesy: D.K. Agencies (P) Ltd. <docinfo@dkagencies.com>

Encyclopaedia of eminent thinkers / K.S. Bharathi.
v. 16-20 cm. — (Gandhian studies and peace research series ; 11)
Vols. 16, 19 by Ashu Pasricha; v. 17-18 & 20 by Jai Narain Sharma.
Includes bibliographical references.
Includes indexes.
ISBN 13: 9788170226840 (set)
ISBN 10: 8170226848 (set)

1. Statesmen—India. 2. Political science—India. 3. India—Politics and government. I. Bharathi, K. S. (Kuduva Swamy), 1956- II. Pasricha, Ashu, 1967- III. Sharma, Jai Narain, 1951- IV. Series: Gandhian studies and peace research series ; 11.

DDC 954.035 22

ISBN -13: 978-81-7022-684-0 (Series)
ISBN -10: 81-7022-684-8 (Series)
ISBN -13: 978-81-8069-496-7 (Vol. XVI)
ISBN -10: 81-8069-496-8 (Vol. XVI)

First Published 2008

Published and Printed by
Ashok Kumar Mittal
Concept Publishing Company
A/15-16, Commercial Block, Mohan Garden
New Delhi-110059 (India)
Phones: 25351460, 25351794
Fax: 091-11-25357103
E-mail: publishing@conceptpub.com

DEDICATED
TO
SUDHA LONNON, DAYRL LONNON
AND
MAYA ISABEL LONNON

PREFACE

Netaji Subhas Chandra Bose was not only born patriotic leader but was also an original thinker. Unfortunately, Indian academics hardly acknowledged this fact and usually take a casual interest in studying this greatman and the principles he stood for, who believed in ushering a new decolonised, united socialist India.

There is no denying of the fact that Subhas Chandra Bose was not a political philosopher in the traditional sense of the term, for he never indulged in philosophical discourses with theoretical premises. He was deeply involved in the national politics of colonial India and concentrated all his energy in overthrowing the British rule. The achievement of freedom was his prime goal. Hence, his political ideas mainly revolved around national freedom which aimed at paving the way for bright future of India. However, there is a spontaneity in his political thinking which cannot be separated from his hectic political life.

As a zero sum player Netaji Subhas Chandra Bose once bluntly said, "I am an extremist and my principle is—all or none". As a genuine leader he had proved how to sacrifice everything for the sake of attaining one of his biggest and cherished goals in life, that is, to liberate the Motherland India from the clutches of British Imperialism. He was not interested in just negotiation for transfer of power from the British hand to Indian hands by breaking the country into two parts: India and Pakistan. He was aware that some leaders were interested to act as "power brokers" between them and the mischievous colonist. While fighting the British, Subhas clearly stated that Britain is India's enemy and Britain's enemy is India's friend. It was his strong sense of nationalism that distinguished him from others. Mahatma Gandhi called him "patriot of patriots". The Mahatma and Subhas were two poles apart in philosophy but each saw the Patriotic soul in the other. It was a tribute well paid. In fact, researcher of this volume feel that Subhas' vast knowledge of the world politics and deeper understanding of the then changing world war related political situation was unique in many ways. Getting international support for India's freedom from the imperialist Britain through military means was unparalleled in Indian history.

In this array of few outstanding men, Subhas Chandra Bose stands out unique as being the modern visionary, whose ideas remain contemporary and valid in free India, even half a century later. Bose was not only clear in his conceptions, but was unfaltering and straight forward in his thoughts. Though academically disciplined in Vedanta and Philosophy, his was a pragmatic mind with strong scientific logic. There was no wavering, no shifting of feet for momentary gain. That is why he stands out being the tallest of them all.

Personal acknowledgments are no easy matters where so many have been kind. My special thanks are due to Professor R.N. Pal, Professor N. Radhakrishnan, Dr. K.S. Bharathi, Dr. A. D. Mishra, Professor Jai Narain Sharma, Dr. Meena Dutta, Dr. Seema Pasricha, Dr. Rashmi Sharma, Mr. Ajay Pasricha (advocate) and Master Hemant Malhotra. Beyond that I can make only an all embracing salute to the galaxy of the scholars whose works I have consulted.

I am also grateful to Shri Ashok Mittal my publisher for his consistent zeal in the printing and publication of this volume.

Chandigarh **ASHU PASRICHA**

CONTENTS

Preface 7

Details of published volume in the series 10

1. Life and Times 11
2. Ideology and Perception 24
3. Freedom and Nationalism 40
4. Concept of Socialism 62
5. Communal Amity and Harmony 79
6. On Fascism 88
7. A Critique of Gandhian Leadership 104

Bibliography 122

Index 133

Encyclopaedia of Eminent Thinkers Series

Vol. I. The Political Thought of **Mahatma Gandhi**/ *K.S. Bharti*
Vol. II. The Political Thought of **Nehru**/ *K.S. Bharti*
Vol. III. The Political Thought of **Rabindra Nath Tagore**/ *K.S. Bharti*
Vol. IV. The Political Thought of **Radhakrishnan**/ *K.S. Bharti*
Vol. V. The Political Thought of **Vinoba**/ *K.S. Bharti*
Vol. VI. The Political Thought of **Aurobindo**/ *K.S. Bharti*
Vol. VII. The Political Thought of **Jayaprakash Narayan**/ *K.S. Bharti*
Vol. VIII. The Political Thought of **Vivekananda**/ *K.S. Bharti*
Vol. IX. The Political Thought of **Ambedkar**/ *K.S. Bharti*
Vol. X. The Political Thought of **M.N. Roy**/ *K.S. Bharti*
Vol. XI. The Political Thought of **Dadabhai Naoroji**/ *Ashu Pasricha*
Vol. XII. The Political Thought of **Veer Savarkar**/ *Jai Narain Sharma*
Vol. XIII. The Political Thought of **M.A. Jinnah**/ *Jai Narain Sharma*
Vol. XIV. The Political Thought of **Tej Bahadur Sapru**/ *Ashu Pasricha*
Vol. XV. The Political Thought of **C. Rajagopalachari**/ *Ashu Pasricha*
Vol. XVI. The Political Thought of **Subhas Chandra Bose**/ *Ashu Pasricha*
Vol. XVII. The Political Thought of **Mohammad Iqbal**/ *Jai Narain Sharma*
Vol. XVIII. The Political Thought of **K.B. Hedgewar**/ *Jai Narain Sharma*
Vol. XIX. The Political Thought of **Bhagat Singh**/ *Ashu Pasricha*
Vol. XX. The Political Thought of **M.S. Golwalkar**/ *Jai Narain Sharma*

1

LIFE AND TIMES

Subhas Bose's father Janakinath Bose had migrated to Orissa in the eighties of the nineteenth century. He had settled down at Cuttack as a lawyer. Subhas was born there on Saturday, the 23rd January, 1897. His father was a descendant of the Bose's of Mahinagar, while his mother Prabhabati belonged to the family of the Dutt's of Hatkhola. Subhas was the sixth son and the ninth child of his parents. He belonged to a well-to-do middle class family and thus had no personal experience of poverty.

Subhas used to feel like a thoroughly insignificant being. His parents loved him. His father had a cloak of reserve around him and kept his children at a distance. Due to his professional work and public duties, he didn't have much time for his family. The mother of Subhas dominated the domestic scene. The presence of so many elder brothers and sisters relegated him into insignificance. However, the family environment helped to broaden the mind of young Subhas.[1]

Subhas has traced the history of his family for about 27 generations. The Boses[2] are Kayastha by caste.[3] The Founder of the Bose family was one Dasarath Bose. He had two sons, Krishna and Parma. Parma went over to East Bengal and settled there. Krishna lived in West Bengal. One of the great-great grandsons of Dasarath Bose was Mukti Bose who resided at Mahinagar, a village about 14 miles to the south of Calcutta. It is because of its association with Mahinagar, that the family is now known as the Bose of Mahinagar. Eleventh in descent from Dasarath was Mahipati. He was a man of outstanding ability and intelligence. He attracted the attention of the then Muslim King of Bengal who appointed him as Minister of Finance and War, and conferred on him the title of Subuddhi Khan. Mahipati was also given a jagir as landed property. The village of Subuddhipur was probably his jagir. It is not far from Mahinagar. Of Mahipati's sons, Ishan Khan was the fourth. He rose to eminence and received titles from the king. The second son Gopinath Bose was appointed Finance Minister and Naval Commander by the then King, Sultan Hussain Shah (1493-1519).[4] He was rewarded with the title of Purandar Khan. He was also given a jagir now known as Purandarpur.

The floods forced a large section of the people to migrate to other places. One branch of the Bose family moved to the adjoining village of Kodalia. This neighbourhood leapt into activity during the early decades of the nineteenth century and produced a number of scholars like Pandit Ananda Chandra Vedantabageesb, Pandit Dwarkanath Vidyabhushan, Pandit Shivnath Shastri, Bharat Chandra Shiromani, Kali Kumar Chakravarty, Kaliprasanna Bose, Harikumar Chakravarti, Satkari Banerjee and M.N. Roy.

Janakinath Bose, the father of Subhas Bose was the thirteenth from Purandar Khan and twenty sixth from Dasratha Bose. Subhas's grandfather, Harnath Bose had four sons—Jadunath, Kedarnath, Devendranath, and Janakinath (Bose's father).

Though by tradition the Bose family was Shaktal, Harnaffi was a pious and devoted Vaishnava. His four sons migrated to different places in search of a career.

Subhas's father Janakinath Bose was born on 28th May, 1860 and his mother in 1869. After passing the Matriculation examination from the Albert School, Calcutta, he studied for some time at the St. Xavier's College and the Scottish Church College. He then went to Cuttack and graduated from the Ravenshaw College. He returned to Calcutta and took his law degree. During the period he came into contact with some prominent personalities of Brahmo Samaj. In 1885 Janakinath Bose went to Cuttack and joined the Bar.

The mother of Subhas Bose belonged to the family of the Dutt's of Hatkhola. The Dutts had attained a great deal of prominence by virtue of their wealth and their intelligence.

Bengal was the first province to come under the British rule. Hence, the visible changes were more visible there than in other places. With the overthrow of the Muslim government the feudal aristocracy lost its importance. It's place was taken by educated middle class Bengalis who could help the new rulers in trade and administration. Most of the prominent personalities that arose soon after the British conquest were Hindus. The most outstanding of them was Raja Ram Mohan Roy (1772-1833) who founded the Brahmo Samaj and was instrumental in introducing the English Education and initiating a number of social reforms. Devendranath Tagore (1818-1905), father of Rabindranath Tagore and Keshav Chandra Sen (1838-1884), Ishwar Chandra Vidyasagar (1821-1891), Ramakrishna Paramhansa (1834-1886) and Swami Vivekananda (1863-1902) were other notable personalities of the time.

Subhas was the sixth son and ninth child in the large family of eight sons and six daughters of Janakinath Bose couple. An interesting point worth mentioning here is that Bose was born 40 years after the first war of independence and when he was 40 years of age he presided over the AICC and blew the conch to declare a war against the British for complete independence.

Like any other typical orthodox Hindu mother, Mrs. Prabhavati Bose filled the hearts of her children with love and regard for the rich traditions and cultural heritage of this great country. In 1902, at the age of five, Subhas was admitted to a school run by the Baptist Mission in Cuttack. The school was meant for European and Anglo-Indian children and only 15 per cent of the seats were given to the children of the Indian aristocracy. Apart from studies, the children were graded for their conduct, neatness, discipline and punctuality.

Subhas maintained satisfactory record in all the gradings but did not like the surroundings and failed to cultivate any friends. However, studies in a Mission school enabled Subhas to be far ahead of the students of Indian schools as far as his knowledge of the English language was concerned. But, the position held by his father was not taken note of either by the teachers or the students in the school. As normally happened in Indian schools, he was not given any preference whatsoever. The Indian students were also barred from sitting in the competitive examinations for scholarships.

They were not eligible to join the Voluntary Corps, while the European and Anglo Indian children were allowed to shoulder the rifles.

Subhas spent seven years in the Mission School before he was qualified to go for higher education. He left the school without regret and without leaving any friends behind and no cause to remember the Head Master or any of his teachers. But he certainly left the school with a feeling that the Indian students, irrespective of their parentage and merit, were discriminated against by the Europeans. Seven years earlier when he joined the school he was an innocent child with a pious mind.

In 1909 Bose joined the Ravenshaw Collegiate School in town where his parentage and the position of his father counted, and it gave him added esteem before fellow pupils and teachers alike. However, he suffered a handicap after he joined the collegiate High School, as he could not learn his mother tongue Bengali and Sanskrit in the Mission School. As a result, he had to learn Sanskrit and Bengali afresh to prepare for the matriculation examination. Lack of knowledge of his mother tongue occasionally brought him insults in the class.

He took to his studies seriously and got over the language problem, but was faced with another problem—sports. Normally, students after returning home from school have snacks and go to play. However, Subhas's parents were against their wards going out for sports in the evenings. The ethical verses in Sanskrit taught Subhas that highest virtue consisted in obeying one's parents. In order to please his parents he took to gardening at home. He helped the servant in tending and watering the plants or in the digging work within the premises of their house or in their adjacent garden. This made him love the plants and consequently enjoy nature's beauty.

The very first day after Subhas joined the high school, he ran into the Principal, Babu Beni Madhav Das, who was on his routine round in the school. He developed in himself a feeling of adoration for the Principal at first sight. Said Subhas: "There was a man who was not an ordinary teacher, who stood apart from and above the rest of his tribe." He eagerly waited to listen to his lecture in the class room. Normally, in any school the Head Master would handle only the higher classes. Therefore Bose began to long for the day which arrived two years later. But his good fortune did not last long. After a few months, orders for the Principal's transfer came. "However, before he left us he had succeeded in rousing in me a vague perception of moral values, an inchoate feeling that in human life moral values should count more than anything else."[5]

As generally happens in the life of any higher secondary student, the number of friends of Subhas progressively increased. He acquired among them a small group as his close circle. Either in the school or outside this group he was deeply involved in discussions on the teachings of Paramahamsa and on the renunciation he taught. His behaviour became peculiar and he was going out of home with his friends for long walks and excursions. At school, none of the fellow students and teachers found fault with Bose or his friends as every one of them was doing well in his studies.

It was during this period that Subhas Bose was drawn towards the works of Swami Vivekananda. He was thrilled to read them. Vivekananda gave him an ideal. The ideal was a work for your own salvation and for the service of humanity. The service of humanity included the service of one's country. Swamiji's biographer and his chief disciple Sister Nivedita, had pointed out, "The queen of his adoration was his motherland. There was not a cry within her shores that did not find in him a responsive echo." Swamiji himself had said, "Say brothers at the top of your voice, the naked Indian, the illiterate Indian, the Brahman Indian, the Pariah Indian is my brother." Talking of the future, Swamiji had remarked that the *Brahman* (religious caste), the *Kshatriya* (warrior caste) and the *Vaisya* (trader caste) each had had their day, and now came the turn of the *Sudras*, the down trodden masses. Vivekananda gave a new interpretation to the ancient scriptures. According to him the real message of the *Upanishads* was strength. To some idle monks he had turned round and said, "Salvation will come through football and not through the *Gita*".[6]

Subhas was barely fifteen when Vivekananda entered his life. Then there followed a revolution which changed his life. Studies began to lose their importance for him. He read a lot on *Brahmacharya* or sense control, meditation, and yoga. He started yogic practices and develop concentration. He was also attracted towards the monks who visited Cuttack from time-to-time. Subhas also engaged himself in village reconstruction at this stage.

Uptill 1911 Subhas was politically underdeveloped. The first political impetus he received was in 1912 from a student about the same age as himself. The student talked to Subhas and his friends passionately about their duty to the motherland. As Subhas approached the end of his school career, the religious impulse began to grow in intensity.

Subhas sat for the Matriculation Examination in March, 1913 and came out second in the whole university. His parents were delighted and he was packed off to Calcutta for further studies.

Presidency College

Subhas Bose had earlier visited Calcutta a number of times. The city had intrigued and bewildered him beyond measures. Subhas had loved to roam about its wide streets and among its gardens and museums. He had felt that one could not see enough of it. It was like a "Leviathan" which one could look at from outside and go on admiring unceasingly.

Though Subhas was passing through a period of stormy transition when he left school, he had by then made certain definite decision for himself. He was not going to follow the beaten track, come what may; he was going to lead a life conducive to his spiritual welfare and the uplift of humanity; he was going to make a profound study of philosophy so that he could solve the fundamental problems of life. In practical life he was going to emulate Ramakrishna and Vivekananda as far as possible. In any case, he was not going in for worldly career. This was the outlook with which he faced a new chapter in his life. Subhas had been greatly benefited by his early interest in religions and his dabbling in yoga. He had learnt to take life seriously. Standing on the threshold of his college career, he felt convinced that life had a meaning and a purpose to fulfil that purpose, a regular schooling of the body and the mind was necessary. In the year 1913 when Subhas joined college, it was almost a fixed idea with him that conquest of sex was essential to spiritual progress and that without spiritual uplift, human life had little or no value.

Subhas joined the Presidency College, which was regarded as the premier college of the Calcutta University. He was sixteen and a half years old when he walked into the precincts of Presidency College. The college at that time had four major groups. There was firstly a group consisting of the sons of Rajas and rich folks and those who preferred to hobnob with them. They dressed well and took a dilettante interest in studies.

Then there was a group of book-worms, well meaning goody goody boys with sallow faces and thick glasses. Thirdly, there was a group consisting of earnest boys who considered themselves the spiritual heirs of Ramakrishna and Vivekananda. Last but not least, there existed a secret group of revolutionaries about whose existence most of the students were unaware. For the first two years of his college life Subhas was greatly under the influence of the group following the teachings of Ramakrishna and Vivekananda. The group followed generally the message of Ramakrishna and Vivekananda but emphasised social service as a means of spiritual development. The main aim of the group was to bring about synthesis between religion and nationalism. The members of the group were exceedingly alert and active. The activity of the group manifested in three decisions. There was a thirst for new ideas. Members of the group were active in recruiting new members. They were also active in developing contacts with the prominent personalities of the day. In 1914 Subhas Bose along with some members of the group visited the poet Rabindranath Tagore who gave a discourse on village reconstruction.

Subhas Bose was deeply impressed by Aurobindo Ghose during his undergraduate days. Ghose was the most popular leader of Bengal despite his voluntary exile and absence from politics since 1909. Surendranath Banerjee was another personality who had made a deep impact on the politics of Bengal. Subhas Bose also went out of his way to get acquainted with such Professor as Sir P.C. Ray, the eminent chemist and philanthropist.[7]

In the summer vacation of 1914 Subhas went on a pilgrimage with one of his friends and visited well known places of pilgrimage in upper India—Hardwar, Mathura, Benaras, Gaya. During this tour, he also visited some places of historical interest like Agra and Delhi. When Subhas returned home, he fell sick, and took some time to recover. It was during this period that the First World War broke out.

During his Presidency College days Subhas often met important men in the terrorist revolutionary movement. However he was never drawn towards them because he believed that the ultimate solution of India would come through the process of national reconstruction.

Subhas was shocked to see the behaviour of the Britishers in Calcutta. Britishers using tram-cars were purposely rude and offensive to Indians in various ways. Sometime they would put their feet up on the front seats so that their shoes would touch the bodies of Indians. Poor clerks going to office would put up with this insult. But it was difficult for others to do so. Subhas was not accustomed to pocket such an insult. Often hot words would exchange between Subhas and the Britishers.

Subhas appeared at the Intermediate examination in 1915 and was placed in the first division. However, he was low down in the list. He had a feeling of remorse and resolved to make good at the degree examination.

For his degree, Subhas took the Honours course in Philosophy. He found interest in studies but was disappointed. He had expected to acquire wisdom. But he actually acquired intellectual discipline and a critical frame of mind.

In January, 1916 Subhas organised a successful strike in the college against the misbehaviour of an English Professor. He was expelled from the college. He tried to get admission in some other college but was refused. A committee of enquiry was set up with Sir Ashutosh Mukherji as chairman. The committee examined Subhas and found him guilty. The elder brothers of Subhas thought that his stay in Calcutta was risky. As a result he was packed off to Cuttack. The tragic events of 1916 had put an end to Subhas's educational career but Subhas had developed self-confidence as well as initiative. He had a foretaste of leadership and of the martyrdom. He had acquired character and was ready to face the future with equanimity.

It was the end of March, 1916 when Subhas came down to Cuttack as a rusticated student. He had the sympathy of his family members. He put his books aside and plunged himself in social service. Subhas nursed the patients and founded a youth organisation with different departments for the physical, intellectual and moral development of the members. He went out on excursions with friends to different places of religious and historical importance. Subhas also used religious festivals for developing group life. During this period he made distinct progress in the practice of self analysis.[8]

Academic studies meanwhile made little progress. Bose's father, who had not been unsympathetic, had nevertheless refused to let him continue elsewhere, insisting that his course in Calcutta be first completed. He was readmitted to the University in July 1917. There was no further setback. He joined the University Training Corps, spending many happy days in it during 1918, became a member of the Students' Consultative Committee and in 1919 got first class in philosophy (Hons.). It was apparent that academically nothing could stand in his way. To his father it was also clear that the nationalist turbulence of Bengal at this time would tempt him sorely: let him therefore go to England and sit for the Indian Civil Service: let power and dignity cure him. The father's decision was sudden and the son's dilemma acute. Does not power corrupt? If he succeeded, what of his nationalism? Yet he had already refused to join the terrorists. He knew the people as many of them were of his own caste and kind but he felt that the time was not ripe for violent revolution.

With many misgivings he agreed to go. This was no change of heart: to enter the Indian Civil Service was still consistent with Indian Nationalism, and the examination would serve him as a final test of his superiority over the European. His over-sensitivity easily found for him further examples of racial prejudice in England. He could still write from Cambridge in November 1919: 'what gives me the greatest joy is to watch the white skins serving me and cleaning my shoes.'[9] This racial complex lost its edge with time, it was on occasion overcome, but its continuing influence was the root cause of the arrogance which he was always apt to display towards the British.

At Cambridge Bose began to make English friends, to enjoy the freedom and the give and take of the Union. In spite of himself he was impressed: 'People here have a sense of time. Many are their defects, but one must bow one's head to their merits.' He admired their energy, love of discipline and *esprit de corps*. Deliberately, and with the single minded concentration that was now the most obvious characteristic of this strange young man, he set out to make himself personally acceptable to them so that he could study them better. The manners of his fellow Indians made him blush: 'Let your one ambition,' he would tell them 'be to leave behind an impression here of flawless spruceness.' And to his friend, Dilip Kumar Roy, in outraged protest: 'Don't for God's sake fling your arms about like lassos!'[10] The freedom of gesture natural to an Indian was, he saw, one cause of the English assertion he resented most of all that Indians were not fit, not civilised enough, for self rule. Always shy with women, he regarded feminine society as a danger to be shunned; he thought he had mastered sex, and he solemnly disapproved of the behaviour of the less restrained. But censoriousness won him no friends; his countrymen called him a humourless prig, and he retaliated by ignoring the Indian student community. He became more concentrated than ever: his methodical work, his abstinence and scrupulous personal discipline made him the envy of his compatriots.

There was not, of course, a sense of humour: yet it was possible for him occasionally to laugh and relax. Years later he told how Mr. Roy had imparted to him some appreciation of music, and nourished his sense of laughter and beauty. But Roy had to drag him from his books: because of the atmosphere of high seriousness in which he had been brought up 'he needed laughter more than the likes of us'.[11] He was also able, exceptionally, to forget his shyness of women: he sometimes stayed in England with Dr. Dharamvir, whose wife was English; she was perhaps in a better position than most to realise and cope with his racial difficulties, and they became good friends.

The harsher qualities were those of success. Bose passed the Civil Service open competitive examination brilliantly in 1920, attaining fourth place, a considerable achievement for one who had been in England barely eight months. But the direction of Indian nationalism was now changing. When Bose set out for England, Mr. Gandhi had not thrown himself into unqualified opposition to British rule. The tragic massacre of Amritsar had only just taken place, and its full horror was still obscured by the seriousness of the disturbances which had occasioned it. In 1920, when, despite official action against General Dyer and the repudiation of his methods by the House of Commons, the English people as a whole appeared to support what he had done, Mr. Gandhi turned away. The Indian National Congress adopted his programme of non-cooperation and his call to Indians to renounce their British titles and appointments.

Subhash debated the matter with his family, particularly with his brother Sarat; even Sarat urged him not to resign life in the service would be tolerable, there was honourable work to be done, the British temper was changing. But to Subhas they seemed to miss the point: 'I must either chuck this rotten service and dedicate myself whole heartedly to the country's cause, or I must bid adieu to all my ideals and aspirations'. No Indian had yet resigned from the Indian Civil Service. If he failed here he would never respect himself again; nothing else was worth considering; he did not intend to marry and had 'at heart no worldly ambitions'.[12] He made up his mind in January 1921: he would offer himself for work at the Congress National College in Calcutta, and on the new nationalist paper Swaraj. He resigned from the Indian Civil Service in April.

The Under-Secretary of State for India sent for him. 'I told him,' said Bose later, 'that I did not think one could be loyal to the British Raj and yet serve India honestly, heart and soul.'[13] His father was grieved, but he would not be diverted: 'If we build our ideals thinking first and last of our family happiness, won't the ideals be wonderful!'

He returned to India in July 1921; to throw himself at the feet of Mr. Gandhi. Gandhi had just made his promise of independence for India within a year, and Bose had put forth his own interpretation regarding the issue. A few months earlier he had talked with Mr. Roy about the possibilities of revolution in India. Indian Revolution, he had said, was 'full-grown when the Russians were learning the trade: even now things were being hatched ... only this time the birds will not be let out until they are full fledged. And then there will be such music at that new dawn'.[14] His face flushed with excitement. And you call them failures? Can't you imagine the beauty of it all—the appalling beauty of the courage of an infant organisation of a tiny handful which dared to throw down the gauntlet to the mighty British. Bose was impatient for battle and thought that this was what Gandhi offered.

His return, then, was as a 'would-be' revolutionary, eager to be called upon for some deed of daring that would justify, to himself and to his family, his spectacular sacrifice of orthodox brilliance. He found Gandhi in Bombay, and the fascination he experienced can be seen in his own account of that first interview:

"I remember vividly the scene of that afternoon... I was ushered into a room governed with Indian carpets. Almost in the centre, facing the door, sat the Mahatma, surrounded by some of his closest followers. All were clad in home made khadi. As I entered the room I felt somewhat out of place in my foreign costume and could not help apologising for it. The Mahatma received me with his characteristic hearty smile and soon put me at ease, and the conversation started at once. I desired to obtain a clear understanding of the details, the successive stages of his plan, leading on step-by-step to the ultimate seizure of power from the foreign bureaucracy. To that end I began to heap question upon question and the Mahatma replied with his habitual patience."[15]

But instead of action, he found non-violence; instead of a revolutionary plan, the pious hope that the British would be 'converted' by Gandhi's methods. Bose left him in disappointment, convinced that he had no 'clear idea of the successive stages of the campaign which would bring India to her cherished goal of freedom.' He went home to Calcutta to work under C.R. Das, the Bengali patriot who had given up his lucrative legal practice for poverty and nationalist agitation. With Das he was to learn many lessons.

REFERENCES

1. Bose, Subhas Chandra, *An Indian Pilgrim: An Unfinished Autobiography and Collected Letters* (Bombay: Asia Publishing House), 1965, p. 3.
2. The original form in Sanskrit is Basu or rather Vasu. In common parlance in Bengali; Vasu has become Bose.
3. The Kayasthas claim to be none other than Kshatriyas or the warrior caste. They are classified among the higher castes.
4. Sultan Hussain Shah was the king of Bengal from AD 1493-1519. He was the first Chief Minister of Sultan Shamsuddin of Bengal. Shamsuddin was deposed and killed by the nobles. Following Shamsuddin's death Hussain Shah became the ruler. He founded the Hussain Shahi dynasty which ruled in Bengal from 1493 to 1539.
5. Adiraju, V.R., *Netaji Subhas Bose: The Great Revolutionary* (Hyderabad: Om See Satya Publications), 2001, p. 47.
6. Gupta, V.P. and Mohini Gupta, *The Quest for Freedom—A Study of Netaji Subhas Chandra Bose* (New Delhi: Radha Publications), 1998, p. 6.
7. P.C. Ray (1861-1944) was born in a zamindar family and received his education at the University of Edinburgh. After obtaining the D.Sc degree in Inorganic Chemistry from aboard, P.C. Ray joined the educational service but he continued with his research in Chemistry. His work lay in the field of nitrates for which he was awarded international recognition. Acharya Ray established the Indian School of Chemistry and the Indian Chemical Society. He was knighted by the British Government and was elected on Honorary Fellow of the London Chemical Society. The Bengal Chemical and Pharmaceutical works is a major industrial organisation set up by him. His book History of Hindu Chemistry is a classic.
8. Subhas indulged in this practice throughout his life. It consisted in the throwing of a powerful searchlight on your mind with a view of knowing yourself better. Usually before going to sleep or in the early morning, he would spend some time over this: the analysis of the whole life, from the former he would get to know more about his hidden desires and impulses, ideals and aspirations. From the latter he would begin to comprehend his life better, to realise his errors of the past and thereby draw conclusions for the future.
9. Roy, Dilip Kumar, *The Subhash I Knew* (Bombay: Asia Publishing House), 1946, p. 53.
10. *Ibid.*, p. 52.
11. *Ibid.*, p. 89.
12. Bose, Subhas Chandra, *An Indian Pilgrim: An Unfinished Autobiography and Collected Letters, op. cit.*, p. 128.
13. Roy, Dilip Kumar, *The Subhash I Knew, op. cit.*, pp. 71-72.
14. *Ibid.*, p. 201.
15. Bose, Subhas Chandra, *The Indian Struggle* 1920-1934 Part II (Calcutta: Published for Netaji Publishing Society), 1948.

2

IDEOLOGY AND PERCEPTION

In October 1924, Subhas Bose, then Chief Executive Officer of the Calcutta Municipal Corporation, handpicked for the post who had been elected mayor on the swarajist ticket, was arrested under the abominable Regulation III of 1818 and later taken to Burma Mandalay jail. He was a prisoner without trial. Among the front ran Congress leaders. After the detention of Lala Lajpat Rai for a year in Mandalay Jail, Subhas alone was the victim of this lawless law, not once but thrice in 1924, in 1932 and again in 1936, the second time when he was the elected mayor of Calcutta. Lala Lajpat Rai and Sardar Ajit Singh were released from Mandalay Jail within a year of their arrest. They had been arrested in May-June 1907. Subhas, a Regulation III prisoner, was not released until the spring of 1927 when doctors suspected him to be suffering from tuberculosis. In one of his letters to Dilip Kumar Roy from Mandalay, Subhas wrote: "I cannot say that I would like to stay in a jail ... The whole atmosphere inside a jail tends, if anything, to pervert and dehumanize human being; and I believe this must be true of all jails, more or less...If a man can find sufficient food for contemplation then his incarceration need hardly hurt him much unless his health desert him. But our suffering is not merely spiritual there is the rub the body too has a say in the business, so that even when the spirit is willing the flesh might be weak. Lokamanya Tilak wrote out his commentary on the *Gita* while in prison...But, withal, his premature death was as certainly attributable to his six years' detention in Mandalay jail."[1]

It was during this phase that he reacted strongly to the argument Roy had repeatedly pressed on him that he was being untrue to his own "*swadharma*" which was deeply spiritual by opting for a life wholly devoted to an organized political movement. Dilip urged him to withdraw into seclusion to discover what his true *dharma* was. On another occasion, seven years later, Subhas wrote from a prison in Madras (now Chennai): "It is easy to say 'live in accordance with your *Swadharma*' but so difficult to know what one's *dharma* is. It is there that the help of a Guru becomes so necessary—and even indispensable."[2] But, answering the same point, nearly a year after his deportation to Mandalay, in 1925 Subhas was openly critical of the call for retreat from the struggle to seclusion for a spiritual quest. His misgivings about Sri Aurobindo's withdrawal from the freedom struggle for his sadhana were explicit: "I subscribe to most of what you write about Sri Aurobindo, If not all is a dhyani (a contemplative) and, I feel, goes even deeper than Vivekananda, though I have a profound reverence for the latter. So I agree with you when you say that one may from time-to-time—and on occasion for a long spell—remain withdrawn in silent contemplation in perfect seclusion. But here there is a danger; the active side of a man might get atrophied if he remained cut off for too long from the tides of life and society ... For a variety of reasons our nation has been sliding pauselessly down to the zero line in the sphere of action; so what we badly need today is a double dose of the activist serum, rajas.

"I say ditto to you again when you say that each of us must strive to develop his powers to their fullness ... Not till our inner being, our *swadharma*, has fulfilled itself, shall we have won through to our inalienable right, adhikar, to what I call real service. To put it in the language of Emerson, we must be moulded from within...The artist's *sadhana* is not the same as the activist's, no more than the contemplative's *sadhana* is the same as the savant's though, I think, in the last analysis, the ideals of all are one ... The nature of each must indicate the clue to the path that is his, the path that leads to his self-amelioration and self-expansion ... It is indeed possible that a man may have to lead, during a particular phase of his *sadhana*, a life which looks on the surface like selfishness or ego centrism.

But while he is passing through that phase he must follow the dictates of his own conscience not those of public opinion."[3]

On 2 January 1932, Subhas Bose was again held under Regulation III of 1818, soon after Gandhi's return from the failed Second Round Table Conference in London. Although this signalled another countrywide government effort to suppress the Congress and abort the chance of a resumption of the Civil Disobedience Movement which had been suspended under the Gandhi-Irwin Pact of 1930, Regulation III was used only against Subhas in January, and a month later, to seize his elder brother Sarat Bose, a leader of the Bengal Congress and a renowned barrister of the Calcutta High Court. As Subhas's health did not improve even after detention at Bhowali's sanatorium (where patients suffering from tuberculosis were admitted), he accepted the government's term that he would be released only when he was on board a ship taking him to Europe.

On 23 February 1933, he sailed from Bombay to Vienna via Venice. Mortified at the sufferings being piled on his friend, Dilip Roy again tried to wean the nationalist leader from his unceasing activism. While Subhas was convalescing in Vienna, Roy sent him flowers from Sri Aurobindo as the seer's benediction. Subhas accepted them with joy and, reverence. But when Roy depreciated in a letter Subhas's total absorption in the freedom struggle, to the detriment of his *swadharma*, which was spiritual and mystical, the rebel leader disputed Dilip's irrational faith in a guru in a strong riposte.

"What he wrote amounted to this that he 'despaired' of our country when men of the caliber of Dilip Kumar and Anilbaran (Roy) could plump for blind faith as against enlightened reason", Roy recalled later.[4] Anilbaran Roy (1888-1974), after taking MA degree in Sanskrit and Philosophy was teaching at a college when he joined the Non-Cooperation Movement. He was General Secretary of the Bengal Provincial Congress Committee under C.R. Das, the President. He was in prison with Subhas Bose in 1924. After his release he withdrew from active participation in the freedom struggle to Sri Aurobindo's Pondicherry ashram. His interpretation of Aurobindo's commentary on the *Gita* was acclaimed in several countries.

The severity of the criticism astonished Dilip. Years later, Subhas in a conversation with his old friend regretted the value judgment he had made on the vocation chosen by some whom he would have been happy to see active in the freedom movement. Before Dilip should be accused of misplaced pertinacity, one particular letter Subhas wrote after he had sailed for Europe could easily have created the impression that Subhas was spiritually ready for a change of course in his life. From a ship taking him to Venice, Subhas wrote:

"In one of your letters, you asked me about my attitude towards Shiva—or something to the effect. To be quite frank, I am torn ... between my love for Shiva, Kali and Krishna. Though they are fundamentally one—does prefer one symbolism to another. I have found that my moods vary—and according to my prevalent mood I choose one of the three forms—Shiva, Kali and Krishna. Of these three again, the struggle is between Shiva and Shakti. Shiva, the ideal yogi, has a fascination for me and Kali the mother also makes an appeal to me. You see, of late (i.e., for the last four or five years) I have become a believer in Mantra-Shakti by which I mean that certain Mantras have an inherent Shakti. Prior to that, I had the ordinary rationalistic view, namely, that Mantras are aids to concentration. But my study of Tantra philosophy gradually convinced me that certain Mantras had inherent Shakti—and each mental constitution was fitted for a particular Mantra."[5]

This encouraged Dilip Roy to stress again in his letter to Subhas the yogi in the latter; and probably he regretted somewhat explicitly that the political leader had suppressed the mystic in his inner soul "through an error of egoistic worldly conscience."[6] There was really no extraordinary change reflected in Subhas's letter to his old friend Dilip. His spiritual inquiry had not ceased when he strode forward to head the radical wing of the country's freedom movement. In his teens he had searched desperately for a personal guru but found none whom he could accept as his preceptor. His respect for those who persevered with their spiritual discipline and inquiry had not, however, diminished. Now, in his thirties, he was not prepared, without being disdainful of those who claimed to have supramental spiritual knowledge or experiences, to accept these perceptions, as he was to say later, on trust.

His own spiritual discipline was not subverted by his political role. Indeed it had become an unostentatious part of his life, influencing his political beliefs and his personal conduct in public life. The youthful phase when he was "sadhu hunting" was behind him. He had concentrated all his spiritual strength in the active element in his character, rajas, the element extolled by Vivekananda as the national trait to be cultivated in India's present phase, for the struggle for independence. In this mood Subhas was unwilling to surrender his reasoning to any one of the numerous incarnate gods the gurus were supposed to be, according to orthodox belief enshrined in the well-known Sanskrit invocation of guru benediction.

How could one possibly invest a human guru with the attributes of divinity, he asked Dilip Kumar. The latter had denied that political activism was disinterested action (*nishkama karma*), lauded in the *Gita*, because Subhas, Dilip argued, certainly was deeply interested in the success of his political movement. Taken aback by the strength of Subhas's rebuttal, Dilip Kumar sent his letter from Vienna to Sri Aurobindo who had also taken an interest in the spiritual queries contained in Subhas's letters to Dilip Kumar Roy.

Sri Aurobindo in a letter to Roy answered at some length Subhas's criticism of "guru vad", the blind faith and matching superstitions the belief has spawned in India. The sage, surprisingly, sounded defensive in his reply to Subhas's criticism. Defending those who placed their total faith in the infallibility of their chosen guru, Sri Aurobindo argued that Subhas's "reasoned faith" in the goal he had set before himself and the country was based after all on an opinion which was no more infallible than another opinion holding the country to be true. If one was led by fallible blind faith, another acted on his fallible opinion. "If one can get a knowledge that knows, it is another matter; but so long as we have only an ignorance that argues—even, faith may be a glint from the knowledge that knows, however far off, and meanwhile there is not the slightest doubt that it helps to get things done," Sri Aurobindo wrote. The question of course is, as it was then, how does one see "a glint of knowledge that knows"? Cognitive faculty can only decide on the basis of faith, the validity of which Subhas was placing under the scrutiny of "reasoning". Sri Aurobindo said that insistence upon action Subhas so fervently advocated was "absurd" if one had not the light by which to act. Human intellect and energy alone could not set right the ills of this world. Intellect had to be illumined by a change of consciousness before the basis of life could be discovered.

Yet Aurobindo never repudiated his own political activism before he withdrew from action to devote himself exclusively to yoga; nor did he underestimate the strength of the Gandhi-led nationalist movement he had broken the bonds with political activities. The savant would not have claimed that the nationalist action programme of the first decade of the twentieth century had been informed with greater "knowledge that knows" than what he perceived in Subhas's advocacy of action in furtherance of his objective—Indian independence. Shortly before he wrote to Dilip Roy to stress the weakness in Subhas's line of thought (rejection of faith in gurus as incarnate gods; and criticism of those leaving the freedom struggle to seek refuge in gurus), Sri Aurobindo in October 1932 had reaffirmed that the work he had initiated in the political field "was destined to be carried forward, on lines, I had foreseen, by others, and that the ultimate triumph of the movement I had initiated was sure without my personal action or presence."

Aurobindo wrote: "I may say that I did not leave politics because I felt I could do nothing more there; such an idea was far from me. I came away because I did not want anything to interfere with my yoga and because I got very distinct ideas in the matter ... There was not the least motive of despair or sense of futility behind my withdrawal.[7]

Sri Aurobindo had profound respect for C.R. Das whom he once described as a philosopher leading a political movement. On Das's untimely and sudden death in 1925, Aurobindo gave his estimate of one who had been his friend and a political associate when both were members of the first secret society headed by P. Mitra:

"Chittaranjan's death is a supreme loss. Consummately endowed with political intelligence, constructive imagination, magnetism; a driving force combining a strong will and an uncommon plasticity of mind for vision and tact of the hour, he was the one man after Tilak who could have led India to Swaraj."[8]

Subhas was Das's political heir and close to his mentor as a son might be. When Das died, Subhas was prisoner in Mandalay Jail, heartbroken at his sudden loss and disconsolate as he could not share his grief with anyone in that distant prison. Later, he was in the habit of confiding his personal doubts and political problems to Deshbandhu's widow, Basanti Devi.

Although Subhas's reverence for Sri Aurobindo never diminished, their common attachment to Deshbandhu did not stop them from drifting apart when the two took very differing positions during the Second World War. On 19 September 1940, Sri Aurobindo took an unequivocal stand in favour of the Allies against the Axis Powers, expressing the belief that Britain's victory would result in an "era of peace and union among the nations and a better and more secure world order". Seen by Subhas Bose, the war had climaxed the prolonged crisis of interests between two blocs of imperialist powers. And the conflagration, by weakening the imperialist powers, offered the subject nations like India a rare opportunity to liberate themselves. When Sri Aurobindo gave the Allies his unqualified spiritual benediction, Subhas was preparing for his secret withdrawal from India to establish contract with Soviet Russia to secure active help for India's Freedom struggle.

Sri Aurobindo backed the Cripps proposal in a telegram to Sir Stafford Cripps, then in Delhi, on 31 March, 1942, when Japan had overrun all the Asian colonies, except India, of Britain, France, Holland and America. Subhas felt so strongly that the Cripps proposals would pave the way for India's partition that, for the first time since leaving India, he began to broadcast from the radio networks controlled by German, Italy and Japan, pleading with Congress leaders to reject the Cripps scheme with both hands. Till then Bose had resisted all persuasions by Berlin and Rome to broadcast over Axis controlled radio net-works because the Axis powers had not made any public affirmation of their support for Indian independence. In his early youth, Subhas searched for a guru and did not find one. Had Vivekananda not died so soon, Subhas probably would have gravitated towards him. Subhas was only five years old when Vivekananda passed away. In his later years, Subhas turned his face against guru-vad, faith attributing to a human guru's divine infallibility. While he still felt that the guru could be an invaluable teacher to help one find his *swadharma*, he was unwilling to surrender his reason or intuitive understanding to any "god incarnate" in a human being. Thus, in spite of his deep respect for Sri Aurobindo, Subhas turned away from Pondicherry. That was not his destination; and he mistrusted the philosophy that weaned talented people from the freedom struggle for a life of seclusion.

Had not Vivekananda offered to remain chained to the cycle of birth, suffering and death, if that was the price he had to pay to allay the misery of the only living god he knew, the pariah and the poor of India? About the time when Subhas was born, in January 1897, Vivekananda, landing in Madras, after his mission abroad to explain the basis of the *Sanatana Dharma*, gave India a new gospel of spiritual activism:

"For the next fifty years let all other vain Gods disappear from your minds. This is the only God that is awake: our own race—everywhere His hands, everywhere His feet, everywhere His ears. He covers everything. All other Gods are sleeping. Why should we vainly go after them, when we can worship the God that we see all around us, the Virat? The first of all worships is the worship of the Virat, of those all around us. These are all our Gods—men and animals; and the first Gods we have to worship are our own countrymen."[9]

Subhas Bose worshipped him: "If he had been alive, I would have been at his feet ... How shall I express in words my indebtedness to Sri Ramakrishna and Swami Vivekananda? It is under their sacred influence that my life got first awakened. Like Nivedita, I also regard Ramakrishna and Vivekananda as two aspects of one indivisible personality. If Swamiji had been alive today, he would have been my Guru, that is to say, I would have accepted him as my Master. It is needless to add, however, that as long as I live, I shall be absolutely loyal and devoted to Ramakrishna Vivekananda ... The aspiration for freedom manifested itself in various movements since the time of Ram Mohan Roy ... but it was never expressed in the political sphere ... The idea of complete freedom is manifest only in Ramakrishna Vivekananda towards the end of the nineteenth century. 'Freedom, freedom is the song of the Soul,—this was the message that burst forth from the inner recesses of Swamiji's heart and captivated and almost maddened the entire nation ... Swami Vivekananda ... laid the foundation for true nationalism in India."[10]

Thus, although Vivekananda left this phenomenal world in 1902, he had in Subhas one of his most illustrious and steadfast disciples. In thought and practice, Subhas remained true till the very last days of his tempestuous life to the spiritual discipline he had readily accepted early in his life. This side of his personal life was never on display; and yet, as he once said, it met a pragmatic necessity of his inner being. During the war years in Southeast Asia, Bose was conscious that he had probably missed the strategic opportunities for intervention in India in 1942. But he accepted with equanimity the great surge of loyalty—and also some acts of treachery—he had seen in the Indian National Army and the settled Indian communities in the countries of Southeast Asia. While rushing from one battlefield to another to succour INA soldiers who had been forced to retreat from the Indian borders by Japan's decision, Subhas Chandra Bose, president of the Provisional Government of Free India, would without any display of his religiosity find time for meditation and reading the *Gita*.

"In his (Subhas Bose's) personal luggage was a little leather bag containing rosary, a small *Gita*, and his spare reading glasses. Often when in Singapore he would commune with the missionaries of the Ramakrishna Mission, or, late at night, he would drive to the mission, there change into a priestly silk dhoti, shut himself up in the prayer room, rosary in hand, and spend a couple of hours in meditation."[11]

Dr. Ba Maw, the prime minister of Burma under the British Act of 1935, who led Burma to independence on 1 August 1943 under the Japanese dispension, soon after Subhas Bose's meeting with Japan's prime minister Tojo, observed a striking dream element in the Indian leader. "I have stressed the dream element in Bose because I really believe that constituted the deepest part of him. 'I often have moments', he told me, 'when I would like to give up everything and spend my days in prayer and contemplation. But I must wait till India is liberated' ... He of course had his relaxed moments, but on the whole he was a haunted man and too much of a national image during a national crisis to be able to forget it for long."[12]

What was the role model Subhas cherished? Was it perhaps the one he saw in Vivekananda who had Subhas as his posthumous disciple! In his own words: "Reckless in his sacrifice, unceasing in his activity, boundless in his love, profound and versatile in his wisdom, exuberant in his emotions, merciless in his attacks but yet simple as a child—he was a rare personality in this world of ours."[13] Nature endowed Subhas with some of the qualities he attributed to Vivekananda.

If Vivekananda helped young Subhas find his vocation in life, Chitranjan Das's leadership shaped his political future. Indeed with Deshbandhu C.R. Das (1870-1925) and his wife, Basanti Devi, Subhas Bose's relationship was almost filial. Death suddenly terminated Das's leadership which, had he lived longer, might have changed the course of India's freedom struggle, Deshbandhu died on 16 June 1925 at Darjeeling. From 25 October, 1924 to 10 May, 1927, Bose was kept in detention without trial, most of the time in Mandalay Jail, under two very harsh laws, Regulation III of 1818 and the Bengal Criminal Law Amendment Act. Yet, on Das's death, when Subhas Bose was only twenty eight years old, many thought of the latter as Deshbandhu's political heir. Had Subhas not been deprived of C.R. Das's leadership and association, within four years of his fateful decision to resign from the coveted ICS to court instead the sufferings and privation of the freedom struggle, it is possible that the story of his life, and the developments in India, would have been different. It is possible to conjecture that Subhas's subsequent life would have been less arduous. Subhas wrote in his book, *The Indian Struggle*[14], the deaths of C.R. Das, Lala Lajpat Rai and Pandit Motilal Nehru, by depriving the Congress of three leaders who could think for themselves and act on their thoughts made the movement vulnerable to mistakes which could not be avoided or corrected in the absence of leaders who were intellectually and politically as powerful as Mahatma Gandhi."[15]

Subhas Bose can be best understood, by placing him within these two politico ideological discourses represented by Tagore and Vivekananda. From the very beginning, Subhas' concerns had been of a much more 'rational-modern' type than could be evidenced in the policies of the Gandhi led Congress of that period. Subhas did not raise issues of ethics, morality, but instead concentrated on social and economic issues. Even before joining active politics, in a letter to C.R. Das from Cambridge (2 March, 1921) Subhas had written:

"Our Congress has no distinctive policy regarding Currency and Exchange; neither has the Congress a clear cut policy about labour and factory legislation. Then our Congress has no definite policy about Vagrancy and Poor Relief, and, again the Congress has probably no determined policy as to the type of constitution we are going to have after the attainment of Swaraj."[16]

For Subhas, Gandhi's was a well meaning but undesirable influence on the freedom struggle in India. He wrote in 1934: "The success of the Mahatma has been due to the failure of constitutionalism on the one side and armed revolution on the other. Since the eighties of the last century, the best political brains among the Indian people were engaged in a constitutional fight, in which the most essential qualities were skill in debate and eloquence in speech. In such an environment it is unlikely that the Mahatma would have attained much eminence. With the dawn of the present century people began to lose faith in constitutional methods. New weapons like Swadeshi (revival of national industry) and boycott appeared, and simultaneously the revolutionary movement was born... during the Great War there was an attempt at a revolution. The failure of this attempt... convinced the Indian people that it was no use trying to resort to the method of physical force. The country was groping for a new method and looking for a new leader (and) there sprang up... Mahatma Gandhi... (Who) appeared as a dynamic force (but)...was not too revolutionary for the majority of his countrymen."[17]

The latent difference between Subhas' militancy and Gandhi's moderation was there as early as February 1931, when "Gandhi confidentially remarked to Viceroy Irwin that Subhas was his 'opponent' and capable of 'denouncing' him."[18] As an acknowledged 'opponent' of Gandhi, Subhas was impatient for India's complete independence. He was an 'exuberant critic' of Gandhi and was an 'enthusiastic fighter'. But "neither in political image nor in diplomatic guile could he compete with the Mahatma. Behind the facade of militancy, Subhas was still hovering in search of a distinct political programme."[19]

For Subhas, Gandhi's popularity was a product of the 'unsophisticated'[20] nature of the masses in India. The Mahatma had 'fully exploited the mass psychology of the people' but in doing so he was expecting many of the weak traits in the character of his countrymen, which had accounted for India's downfall, to a large extent.[21]

Subhas was convinced that, it was the very 'spiritualism' of the Indian people which was acting as a barrier to modern scientific development.[22] What was a source of major irritation for Subhas was the consciously anti-modern position taken by Gandhi and his followers. Even in, what he called a 'politically-minded province like Bengal', 'a resolution moved at the Jessore Provincial Conference in 1923 to the effect that the goal of the Congress was not spiritual swaraj but political swaraj, was defeated at the end of a heated debate.' Even the liberal intelligentsia, who were opposed to Gandhian politics, had to bow down before the enthusiastic popular support that the Mahatma managed to garner.[23] On the other hand, Subhas was for scientific social emancipation based on economic self-sufficiency. As a part of his attempt to inculcate a scientific rationalistic spirit within the younger members, he invited Dr. Meghnath Saha to deliver a lecture on economic reconstruction, at the Bengal Youth Conference held in 1922.[24] This was yet another conscious attempt to expand the domain of the 'economic question' within the arena of political debates and to shift it away from the ethical moral influence of a Gandhian type of politics.

In Subhas's perception the secret to Gandhi's success amongst the masses lay in his use of the backward elements of 'the mass psychology of the people'. The fact of Gandhi's ability to appropriate the popular classes had been instrumental in his rise within the bourgeois national movement, and the support he received from the national bourgeoisie.[25] And it is this very popularity of the Mahatma's image, which acted as a barrier to the development of an alternate leadership within the national movement. Subhas, as one such aspiring leader, was constantly faced with Gandhi's obvious superiority in any form of mass mobilisation, and the tremendous moral authority Gandhi wielded, as a mass-leader.

Basically an urban politician, the 'modernist' Subhas's failure to mobilise the 'unsophisticated' Indian masses, sometimes translated into a deep hostility towards mass mobilisation itself. Thus, to Subhas, the means to mass mobilisation was largely through the use of instruments of propaganda, such as the Press, and more importantly, inciting the masses through rumour mongering.[26] Subhas' modernist temperament found propaganda through the Press, an important and desirable political practice. Along with his mentor C.R. Das, Subhas had been stressing on the necessity for propaganda in foreign countries. In his presidential address at the Gaya Congress, C.R. Das stated: "There should be established Congress Agencies in America and in every European Country."[27] Subhas also wrote in 'The Indian Struggle 1920-42,' "There is probably nothing which I admire so much about the Britishers as their skill in propaganda." Bose believed that Indian propaganda abroad was absolutely necessary for National advancement. To reach this goal Subhas did his best to establish the kind of Cultural and Commercial Associations "which might grow, endure and contribute to long-term links."[28]

It was propaganda of the other kind, i.e., 'rumour-mongering', which was Subhas' cause for concern. In the pre-modern mode of communication of the 'rumour', Subhas recognised the importance of modern political forms of organisation. 'Rumours' had always been the greatest source of dread for the colonial state, in its counter-insurgency maneuvers, as it was an unnamable, and uncontrollable force without an identifiable source. It was a 'crime' without a 'criminal'. Subhas' quest for an adequate but modern political form, was naturally inimical to this overtly pre-modern and undisciplined mode of political organisation. It was this uncontrollable character of mass-movement which scared Subhas, and made his relationship with the 'masses' extremely ambivalent.

It is no wonder then, that to Subhas, the most serious problem facing India, appeared to be that of middle class unemployment. Thus he wrote, "Under the present conditions in the country, if the middle class is destroyed, the whole nation would be crippled forever."[29]

Infact he came into political prominence in 1923, as an executive officer of the Calcutta Corporation, under the Mayorship of his mentor Chittaranjan Das. This 'constitutionalist' politics of the Swaraj party, at first sight appears to be contradictory to Subhas's professed 'revolutionary' temperament. However, he continued to have connections with the revolutionary terrorists of Bengal, and was deported to the Mandalay Jail in 1924, for being part of an alleged 'revolutionary conspiracy'.[30] Even in regards to the Swaraj Party's participation in council politics, Subhas argued that it was only through a resistance to the bureaucracy from within the government that the colonial state could be forced to respond to popular demands.[31] Even in 1940, in the Editorials of his 'leftist' Party (Forward Block) organ, 'Forward' he justified the coalition with communal parties like Muslim League and Hindu Mahasabha, to defeat the bureaucracy in the Calcutta Corporation election.[32]

It is perhaps possible to say that it is the active presence of middle class problems in Subhas's which formed the immediate background for his political practice. Despite the concern for rural reconstruction as evidenced in the swarajist notion of *palli- samaj*, a close reading of Subhas' texts reveals his essential involvement with middle class problems. Indeed, considering the overwhelmingly urbanized *'bhadralok'* character of mainstream political leadership in Bengal, it was quite 'natural' that Subhas' immediate 'political world' would consist of essentially middle class problems and aspirations. This does not, however, mean that he was not concerned about the position of the depressed classes. Indeed he was one of the first to start "housing for workers, free primary education and free medical services as Chief Executive Officer of the Calcutta Corporation".[33] Despite that, his letters from Mandalay Jail (1925-27) reveal his overwhelming concern for largely middle class issues. He would regularly send his advice and consultation on many problems of the Calcutta Corporation, most of which were typical everyday problems of the urban middle class, such as pensions, holidays, and provident funds of government employees, compulsory primary education, the opening of central markets in Calcutta, problems of civic amenities, etc. The importance of the middle class in the political ideology of the Swaraj party, is typified in the writings of Sarat Chandra Chattopadhyaya, a close associate of C.R. Das and Subhas Bose. In his only overtly 'political' novel *Pather Dabi* (1926), the hero, Sabyasachi declares that it is the middle-class and not the workers or peasants who could bring about a true revolution in the country.[34] All this reveals a fundamental consistency between Subhas' impatience with Gandhian mass politics, his faith in the progressive role of science, and his belief that it was the middle class which was the most progressive section of Indian society.

REFERENCES

1. Dilip Kumar Roy, *Subhash I Knew, op. cit.*, pp. 207-209.
2. *Ibid.*, pp. 91-92.
3. *Ibid.*, pp. 220-221.
4. *Ibid.*, p. 165.
5. *Ibid.*, p. 98.
6. *Ibid.*, p. 102.
7. Sri Aurobindo, *Sri Aurobindo on Himself* (Pondicherry: Sri Aurobindo Ashram), 1972.
8. *Ibid.*
9. Swami Nikhilananda, *Vivekananda* (Calcutta; Advaita Ashram), p. 231.
10. Swami Jyotiramayananda (ed.), *Vivekananda, With a Garland of Tributes* (Madras: S. Ganeshan), 1988.
11. Ayer, S. A., "Unto Him A Witness", in *Subhas Chandra Bose, The Springing Tiger by Toye, Hugh* (New Delhi: Jaico Publishing House), 2001, pp. 88-89.
12. Ba Maw, *Breakthrough in Burma, Memoirs of a Revolution, 1939-1946* (London: Yale University Press), 1968, p. 349.
13. Swami Jyotiramayananda (ed.), *Vivekananda, With a Garland of Tributes, op. cit.*, p. 328.
14. Bose, Subhas Chandra, *The Indian Struggle 1920-1934, op. cit.*, p. 61.
15. *Ibid.*
16. Bose, Subhash Chandra, *An Indian Pilgrim: An Unfinished Autobiography and Collected Letters* (Bombay: Asia Publishing House), 1965, p. 185.
17. Bose, Subhas Chandra, "The Role of Mahatma Gandhi in Indian History", *The Indian Struggle 1920-1934, op. cit.*, pp. 294-95.

18. Gandhi-Irwin Conversation, Extract from Mahadev Desai's manuscript diary, 19 February 1931, "Gandhi Collected Works, Vol. 45, cited in Amit Kumar Gupta, *A Colony A World War and a Patriot*", in A. C. Guha (ed.), *The Story of Indian Revolution* (Calcutta: Asia Publishing House), 1972, p. 208.
19. Gupta, A.C. (ed.), *Studies in the Bengal Renaissance, in Commemoration of the Birth of Bipin Chandra Pal* (Calcutta: Asia Publishing House), 1958, p. 577.
20. Bose, Subhas Chandra, "The Role of Mahatma Gandhi in Indian History", *The Indian Struggle 1920-1934, op. cit.*, p. 213.
21. *Ibid.*, p. 114.
22. *Ibid.*
23. *Ibid.*
24. Chatterjee, Santimony (ed.), *Collected Works of Meghnath Saha* (Bengali) (Calcutta: Asia Publishing House), 1993.
25. Chatterjee, P., "Gandhi and the Critique of Civil Society" in R. Guha (ed.), *Subaltern Studies III* (Delhi: OUP), 1989, p. 156.
26. Bose, Subhas Chandra, "The Role of Mahatma Gandhi in Indian History", *The Indian Struggle 1920-1934, op. cit.*, pp. 114-5, 200, 207, 232.
27. Sen, Rajen (ed.), *Deshbandhu Chitta Ranjan* (Calcutta: Ananda Publishers), 1927, p. 210.
28. Gordon, Leonard, A., "Ambassador of India in Bondage", in *Brothers Against the Raj* (Delhi: Viking), 1990, p. 180.
29. Bose, S.C., *Samagra Rachanabali*, Vol. 1 (Calcutta: Ananda Publishers), 1980, p. 219.
30. Bose, Subhas Chandra, "The Role of Mahatma Gandhi in Indian History", *The Indian Struggle 1920-1942, op. cit.*, p. 106.
31. *Ibid.*, p. 99.
32. Sarkar, M.L. (ed.), *Subhas Basu 1939-1940* (Bengali), translated by Nripen Chakravarty (Calcutta: Forward Publishing Concern), 1975, pp. 138-40.
33. Ghosh, Chitra, "The Socio-Political Ideas of Subhas Chandra Bose", (Unpublished Paper, Presented in the Seminar of J.N.U. on 10 11 November, 1997), p. 3.
34. Chattopadhyaya, Sarat Chandra, *Sharat Sahitya Samagra*, Vol. 2 (Calcutta: Ananda Publisher), 1986, p. 1241.

3

FREEDOM AND NATIONALISM

The goal of Subhas Chandra's thoughts and ideas was the well being and development of the entire people without bias against any caste, class, sex or religion. Hence, the programme he wrote was universally true for all. He was also not ready for any compromise whether political, social or economic. He also firmly and equivocally believed that political freedom must precede freedom from social and economic bondage. To attain this freedom nothing could deter him. He was himself prepared to make the supreme sacrifice and expected all who followed him to do the same. He further believed that India's freedom will herald the freedom of all interests.[1]

Subhas had decided on his life's role and work from his student days, particularly after he went to England to study and compete for the ICS and the Tripos at Cambridge. In letters wrote to Deshabandhu Chittaranjan Das in 1921 from London, he had written about, what he thought, would be the fields suitable for him to work on his return home, having firmly decided that he will not join the Civil Service and be a servant of the bureaucracy. Believing that education was the primary need and necessity of any and all people, he offered himself as a teacher in the National College set up by Das; or as a journalist in the nationalist papers that the leader thought or planned for. Bose steadfastly believed that the primary need of the Indian people was literacy and employment. Also, that if a movement or a party was to be actively successful, there was the need for adequate and planned publicity and propaganda. These ideas Bose held on to, all his life—there was no shifting of position from basic faith and belief. Coming back to India and joining the Swarajya Party of Deshbandhu Das, when Subhas was appointed to the post of Chief Executive Officer of Calcutta Corporation, one of his first tasks was to start free primary evening and night schools for the depressed classes so that everyone who was forced to be involved in earning their daily bread during the day could learn the basics of education in their spare time. Mayor Chittaranjan Das spoke of the programme on assumption of his office, identifying three basic needs—housing for the workers, free primary education and free medical service.[2] Chittaranjan Das also gave Subhas the responsibility to publish the weekly news magazine of the party (*Banglar Katha*)[3] to publicise the programme and work of his party. Within ten years the number of free primary schools run by Calcutta Corporation increased from 3 to 250, under the able supervision of Khitish Prasad Chattopadhyaya who was brought in by Subhas. Nearly 1000 teachers were trained at Corporation's expense at the Teachers' Training Institute of the Education Department. Free medical centres were set up, where special care was taken for children, who were daily given free milk. The labour leader of England, Brailsford wrote that "the Corporation Schools reached a good standard, far better than they were before".[4]

As Subhas became more and more involved in politics, he developed his ideas of free India on a surer basis. He was a strong believer that a socialist system of state organisation was necessary to ameliorate the evils of the alien system of Government, which extended poverty, illiteracy and ill-health. An alternate system had to be thought of and the blue print prepared. Subhas realised that social emancipation had to be preceded by economic self sufficiency and well-being. The process was realistic and factual. It could not be an idle dream. Therefore, at the Bengal Youth Conference held in 1922, he invited Dr. Meghnath Saha to deliver an address and talk about economic reconstruction. Bose, developed a lasting friendship with the scientist, Dr. Saha, who, also later on, after the Haripura Congress, became a member of the Planning Commission (set up by Bose, President of the Indian National Congress). Dr. Saha advocated eloquently the need for scientific development of large scale basic industries.[5] In 1935, Meghnath Saha started a journal, in the first issue of which he severely criticised the development ideas of Gandhi referring to it as a "policy of charkha and bullock-cart" which will lead India to nowhere! In the October issue of the Saha's journal, *"Science & Culture"*, Subhas Chandra wrote an article, entitled "Some Problems of Nation Building". Both advocated the introduction of Five year Plan system for organised and planned economic development. At a meeting of the Congress Ministers, as also in a meeting of the Working Committee of the Congress, at Bombay in 1938, Bose fervently pleaded for planning to develop a modern industrialised nation, in place of the village economy, which had till then been idolised by the Congress Party. Subhas further developed his ideas and revealed in his Presidential address at the Haripura Congress in 1938.

Later in the same year, Bose convened and addressed a meeting of the Industry Ministers of seven Provincial Congress Ministries where the Congress had won majority, when he told them that side by side with the fight for liberation, they must also plan and think about national reconstruction, so that people may be ready once the country was free. Subhas Chandra voiced the same opinion at the All Party Political Conference, held in London in 1933. Though he was invited to be the President, the British Government did not allow him to enter Britain. Hence his written address was read at the meeting.[6] He also sternly opposed the acceptance of the Dominion status for India, which in his opinion would not be able to eradicate the inhibiting influences of the British Raj on the Indian economy and pleaded for complete independence. He believed that if the economy of a country had to flourish and develop, foreign investment had to be replaced by indigenous investment. However, he held that to put economic improvement and industrial development before India is politically free, was to put the cart before the horse.[7] Political power was a necessity as any comprehensive scheme of industrial development under state ownership and control would be indispensable to take the country out of the mire; but then the power to implement such a policy must be there.

Bose was a staunch Indian nationalist. As a logical sequel to his patriotic zeal, Bose developed an invincible faith in India's destiny, and hence took up the case of India's independence as 'a spiritual goal and a righteous cause'. He was not a believer in any escapist philosophy but a revolutionary who was drawn to politics inspired by the love of his country and mankind.[8] The deification of India as the mother and the glamour attached to sacrifice for the motherland constituted the main trend of his strategic thought, the ultimate policy ambition and contributed to the development of his concept of nationalism on the basis of a religious consciousness. India thus featured in his thought not as a mere territory but as a spiritual being, 'a living thing'.[9]

To attain India's liberation he laid his firm faith on revolutionary struggle, which consisted of two phases: a phase of nationalist struggle against British Imperialism and a phase of inter-class struggle against all privileges, distinctions and vested interests. Bose was equally of the firm view that no country could win its independence without any sacrifice and without fighting a war. Such a conviction led him to advance a militant *strategy of direct action* and made him an ardent advocate of revolution, as against Indian National Congress plea for moderation based on constitutional methods. He exhorted all people to be prepared for *dharmayuddha*, i.e., making sacrifice for a righteous cause, as he viewed patriotism with an added emphasis of religiosity.[10] His strategy of direct action and his emphasis on force and violence for fighting the war of India's liberation was thus in full accord with his strategic outlook. It was this approach which made him a subject of bitter controversy in the then Indian socio-political life.

Secondly, Bose from his own analysis came to the conclusion that all the efforts of the Indians themselves may not be sufficient to expel the British and, therefore, he wanted to rouse international public opinion in favour of India's liberation movement. He was of the view that India must exploit an opportune international situation and seek foreign help to drive out Britain from India. The Second World War, he thought, provided such an opportune moment he was looking for and afforded him to organise an army and equip it with arms and ammunition to launch an armed struggle against British rule in India. It was for this reason Bose escaped from India during the War to seek international help so as to fulfil his determination to win India's freedom.

One may ask, why did Bose postulate such a strategic concept which brought him into confrontation with the Congress hierarchy? He was certain that Britain was unlikely to grant independence to India of its own free will, as many in the Congress hierarchy believed. Therefore, he thought it appropriate not only to resort to a strategy of violence involving direct action but also was legitimate to seek international assistance in support of his strategy to overthrow the British.

It may be mentioned that most of Bose's contemporary revolutionary leaders, including Mao Zedong, Ho Chi Minh, Tito held the same view as far as their perception of the external enemy was concerned, though direct onslaught by the Axis powers on their territory had left them with little option but to join in the Allied counter-attack on a commonly perceived enemy. In the case of India, Bose had a different perception: for unlike China, Vietnam or Yugoslavia, India was not a direct military target of any of the Axis powers.

Bose's strategic posture would thus appear quite rational, commanding, and superb. Having read Clausewitz and being convinced to ultimate victory of his mission, he adopted the slogan of 'total mobilisation for a total war' and his battle cry became *Chalo Delhi* (To Delhi) in Clausewitzian style of a zerosum game. Obviously, India, being under British colonial hegemony, was to emerge as the 'front' in the liberation struggle, while the countries of Southeast Asia, where his forces temporarily camped, were to appear as the 'rear areas' from which the struggle would be spearheaded by mobilizing support of the immigrant population as well as of the foreign powers. Such ideas, though developed independently, were consistent with the revolutionary blue print of his great revolutionary Chinese contemporarian, Mao Zedong, ideas which were later further developed by Ho Chi Minh and Vo Nguyen Giap in the course of the Vietnam War.[11]

Indeed in all these spheres, Bose proved to be quite innovative as he sought to mobilise help from what was in essence an 'enemy territory' and was quite unconventional in nature. His aim was to draft support of a couple of millions of Indian civilians in Southeast Asia and recruit Indian soldiers employed with the British colonial army.

It is true that in the military field Bose had no prior exposure, direct experience or background. He had only relatively small training as a member of the University corps. It was hence suggested that he had an 'incredible naivete in military matters';[12] yet he exacted great loyalty and enjoyed tremendous amount of respect from his adherents and subordinates in the Indian National Army (INA). Somewhat echoing words of Napoleon, whose undaunting spirit to overcome any hurdle inspired him,[13] he complimented his soldiers by telling them: 'conduct yourself in a manner so that your countrymen may bless you and posterity may be proud of you'. While lamenting the odds confronting his forces in the INA, he could still inspire them with confidence and Napoleonic voice, I can offer you nothing for the present except hunger, thirst, privation, forced marches and death', but 'if you follow me ... I shall lead you to victory and freedom'.[14] And, again with Napoleonic spirit, while assuring his soldiers that he will be with them 'in darkness and sunshine, in sorrow and in joy, in suffering and in victory', he was himself 'tireless, taking little sleep and working far into the night' so as to fulfil his promise.[15]

Although a committed revolutionary, Bose was not an anarchist. Nor did he believe in terrorism or violence merely for the sake of it, though he did admire the immense sacrifice rendered by the terrorists for the great cause of India's independence. So long as he was in the Congress he wanted to see it developed into a mass-based political organization, able to articulate and mobilise all the energies of the nation for the attainment of the goal of India's freedom. Hence he firmly believed that India's independence was to be won by the development of the Congress into a parallel government, with each department and instrument of British administration in India to be duplicated by the Congress. He wanted India to be a republic and its future constitution to be republican. Hence he rejected Dominion status as he was concerned that it would merely help the British to perpetuate their imperialist and capitalist interests in India. Indeed, such a stance compelled Gandhi himself, who was inclined to such an idea, to move a resolution in 1929 at the Lahore session of the Congress which defined Congress' objective as 'complete independence'.[16]

Concepts of Bose's Philosophy and Idea of History

During his hectic political life, Subhas evolved his political ideology. He interpreted Indian history by saying that history of India has to be recorded not in decades or in centuries but in thousand years long history. During its long history, India had passed through various vicissitudes of fortune, as neither for the individual nor for the nation it is possible to have an uninterrupted career of progress and prosperity. The same was true for India which has been always characterised by a very high level of culture and civilisation. He summarised his readings of Indian history as follows:

(1) A period of rise has been followed by a period of decline to be followed again by a new upheaval;
(2) The decline was the result chiefly of physical and intellectual fatigue;
(3) The progress and fresh consolidation was brought about by the influx of new ideas and sometimes by an institution of fresh blood;
(4) Every new epoch has been heralded by people possessing greater intellectual power and superior military skill;
(5) Throughout Indian history all foreign elements have always been slowly absorbed by Indian society. The Britishers are the first and the only exception to this; and
(6) Inspite of changes in the Central Government, the people have all along been accustomed to a large measure of real liberty.

To regain the lost glory, Subhas wanted regeneration of India. For that he felt the need of spirit of militant nationalism.[17]

Militant Nationalism and Patriotism

This is essential to rejuvenate India to achieve freedom. Subhas believed in non-violent path and mobilisation of the people for the cause of freedom. That is why he supported non-cooperation and civil disobedience movement. But its withdrawal at the crucial moment of history on the pretext of violence and an opportunistic compromise had been severely criticised by him. He wrote: "If our policy had been one of uncompromising militancy, the Bardoli surrender of 1922 would never have taken place—nor would the Delhi Pact of March 1931—when the situation was opportune". He felt that "Freedom intoxicated missionaries who are morally prepared to undergo maximum sacrifice and suffering are necessary to attain success in the mission."

Bose's Views on Nation Building

For Subhas Bose, a nation is not merely to satisfy narrow selfish ends. Addressing the students of the Central Provinces and Berar at Amravati, he said: "I have already said that we shall have to change some of our existing ideas of values with regard to good and evil."[18] He advocated the need of complete overhauling and recasting of the present stereotyped mode of existence which would lead to true complete national solidarity and would give India a seat of glory amongst the great civilised countries of the world. For him, life, only when it is inspired above the ordinary by some greater and nobler ideals, can have any value or significance worth the name. He asserted that a nation can not exist, nor it has right to exist or live, if it has no ambition. However, he also warned that a nation should not strive for progress to satisfy only narrow selfish ends, rather it should march onwards in order to show that it might effectively make its own contribution towards the growing liberty and greatness of the human society. Thus, it should aim at turning this world into a happier and nobler place for mankind.

Swadeshi and Nationalism

Subhas Bose asserted that 'swadeshi' is necessary for nation building. Nationalism is not only political movement but ethical as well, according to Bose. Its ethical aspect reflects in the adoption of swadeshi which has common cause and could be practised religiously. Because when one buys an indigenous product, even though of worst quality and higher price, he helps the nation.

Swadeshi is better than the protection of the industry with other means as protection etc. cannot give impetus to efficiency in the indigenous industry in its infancy. Thus, swadeshi in combination with sacrifice for nation ensures improvement in the indigenous industry.

The Concept of Freedom

Overthrowing the foreign yoke, is the immediate task towards an ideal of achieving freedom. Bose made this clear while addressing the students at Lahore in October 1929. He said that the ideals we have to uphold is an ideal which will galvanise our whole life. That ideal is freedom. But freedom is a word which has varied connotations. He explained, for him freedom is an all round freedom, i.e., freedom for the individual as well as for the society; freedom for the rich as well as for the poor; freedom for men as well as women; freedom for all individuals and for all classes. This freedom implies not only emancipation from political bondage but also equal distribution of wealth alongwith abolition of caste barriers, social inequities and, destruction of communalism and religious intolerance. Thus for Bose, freedom has many facets and to achieve real freedom, socialism, was necessary.

Bose's concern however, was not only freedom or independence. He held a great vision for India's unity and a plan for free India. He held the view that without obliterating the ingrained and vested evils such as poverty, ignorance, caste system, communalism and other social anachronisms and religious divisiveness India can hardly be free or can be turned into modern democracy. Caste and communalism must not be allowed to hinder a free and democratic India.

Here begins Bose's second phase of Indian revolution, the phase of revolutionary construction at home. As part of the second phase of his struggle, Bose laid wholehearted faith in socialism, as it could form part of a general strategy of involving the masses in the freedom movement and would help eradicate barriers and distinction based on wealth, caste, sex and religion, all of which must go. Thus, in addition to its pragmatic utility, the humanitarian and egalitarian aspects of socialism had great appeal for Bose'.[19] He was in favour of a synthesis and in this, his views were again coloured by national considerations. He was all for social upliftment of women and wanted them to have equality of status with men, but he believed that the latter would never voluntarily concede equality of status. Hence, he felt, that for women there is little option but to struggle to assert their right.[20] He himself created an example to facilitate liberation of women by setting up the Rani of Jhansi Brigade during the War consisting only of women so that the latter could be prepared to play due role in the liberated, independent India.

Bose abhorred the emerging trend of communal politics in India, intentionally well laid by the alien authorities as in the case of the Catholic Protestant question in Ireland.[21] He himself sought to bring together the two communal parties, the Hindu Mahasabha and the Muslim League, in a joint fight with the Congress for India's independence and even suggested to Mohammad Ali Jinnah that in the event of such a joint struggle taking place, the latter could become the first Prime Minister of free India. Perhaps in order to accommodate Muslim sentiments, a special session of the Bengal Provincial Conference of the Congress Forward Bloc, founded by Bose, decided on 25 May 1940 at his instance to observe 3 July 1940 as Sirajuddowla Day, in honour of the last independent Nawab of Bengal. But by that time he was already ousted from the Congress itself, which was perhaps a prelude to his removal 'from the political scene in India and his dream of a united India was belied'.[22]

In this context, it may be recalled that Bose's election in 1938 as Congress President by Gandhi himself, despite his ideological differences with Bose, was based on the premise that Bose, with his 'secular approach' and his acceptability to the Muslims, would be able to arrest the mounting tide of communalism in India. And once elected he assured the Muslims that they had 'nothing to fear in the event of India winning her freedom—on the contrary, they have everything to gain'.[23] It was only natural that his ouster from the Congress and his attempted political diminution deliberately pursued by the Congress leadership had the effect of providing upper hand to the communal forces within and beyond Congress, paving the way for the subsequent partition of India.

To attain his goal, Bose also propounded the doctrine of *Samyavada*, which sought to accommodate the common trails of nationalism, fascism, national socialism, socialism and secularism. The ideals he sought to incorporate included complete national independence and uncompromising anti-imperialist stand, scientific large scale production for the economic regeneration through a modern and socialist state, social ownership and control of both the means of production and distribution, freedom of religious worship and equal rights for individuals, linguistic and cultural autonomy for all sections of the Indian community, application of the principle of equality and social justice in building up the New Order in free India. All these would be enforced through establishing *Samyavada Sangh* in India, a centrally controlled all Indian party, representing all classes of people and committed to uphold the eternal principle of India's political independence, equality and justice.

On basic questions of strategy and tactics, Bose was thus immensely thoughtful and farsighted. For resolving fundamental problems of India's national life he conceived a comprehensive plan with two components: an immediate component of struggle against colonialism and a long-term component of national reconstruction. The immediate component of the plan included preparing the country for self-sacrifice towards the objective of liberation, to keep it unified and to provide scope for local and cultural autonomy. Through elaborate planning and appropriate organisation, the country would be prepared for struggle and self-sacrifice, and the people would have to be drawn to a strategy of direct action against foreign imperialism.

Bose thus proved himself to be a great strategic thinker and a good tactician with a pragmatic vision about the future of India. His credit lies not merely in articulating his ideas and theories but also in seeking to translate his politico-strategic theories into a programme of action. During the Quit India Movement he innovated the concept of 'non-violent guerrilla warfare' to be undertaken within India and made a remarkable contribution to the theory of boycott and passive resistance.[24] His programme of action, designed to give the Indian mass movement new dimensions of operational strategy and effectiveness, included prevention of tax and revenue collection, stay-in-strike by workers to hamper production, social boycott of the British and the pro-British Indian elements, sabotage of the British administration and their military functionaries, creation of secret guerrilla bands for the acts of sabotage and actual efforts to seize power by force of arms.

With his strategy of direct action and theory of boycott, Bose also popularised the tactical concept of mass base for political parties to fight a revolutionary struggle, stressing the importance of closely associating the youth, students, the labouring classes, workers and peasants with the political life of the nation in order to galvanise the national movement.

While in Congress, he held the view that the INC could capture power with an intense nationalist struggle which would culminate in a general strike and civil disobedience. That would paralyse the administration and running of the government would be made impossible. Because the colonial government itself is likely to follow a policy of persecution and repression, the jails would be full, the government itself would be demoralised, the bureaucracy would be forced to yield to the demands and desires of the people and the British authorities could no longer count on the loyalty of their servants. The Congress could then take full command of the situation by developing its organisational structure into a full and parallel government, duplicating each department and every instrument of the British hierarchy of administration. Labour should be organised, political education be provided for the masses, all preparations should follow for capturing power from the British.

However, Bose's embittered relations with the Congress hierarchy over the programme of direct action against the British advocated by him, when the latter had confronted the Axis powers, including the issue of an ultimatum to the British and mobilisation of international support in favour of India's nationalist struggle, led to his resignation and eventual expulsion from the Congress. The machination of the Congress hierarchy together with his frequent persecution by the colonial authorities in India made him desperate and had left him with little option but to seek to redeem his framework of action from abroad.

Bose became President of the Indian National Congress at a very critical juncture of international diplomacy when the Axis powers were on the brink of their forward march. As a pragmatic politician, fully committed to the cause of India's liberation, he was inclined towards playing a *realpolitik* and wanted the Congress to pursue a foreign policy which would be based not on idealism or emotionalism but on considerations of India's national interest. His view was that in the pursuit of foreign policy, Congress should set aside its inhibition about the international politics or conduct of any country or the form of its state/government, as the people of every country, whatever their political ideology, sympathised with the aspiration of the Indian people for freedom and independence. He took the subtle position that the Congress leadership should tone down their utterance against the militarist and expansionist activities of the Axis powers and should take a leaf out of diplomacy of the USSR, which did not hesitate to make alliances with non-socialist states as long as such an alliance served Soviet interests. In fact, apart from the question of the pace of the nationalist movement, and appropriate strategy and tactics to be followed, one of the main reasons of Bose's suspension from all executive positions in 1939 was his differences with the Congress hierarchy over the war issue and his attitude towards world affairs.

Bose's 'whole life', as he stated, 'is one long, persistent, uncompromising struggle against British imperialism, and is the best guarantee of my *bona fides*'.[25] He himself saw his leanings towards the Axis alliance as a 'hazardous mission' and his sole object was the expulsion of the British from India; in this case, if he had any triumph he would share that with Gandhiji and the Indian people.[26] But Bose was not an Axis apologist and was not certainly soft toward the expansionist manoeuvers of the Axis powers. Rather it was under his stewardship as President, the Congress had decided to observe 12 June 1939 as 'China Day' throughout India in solidarity with the Chinese people fighting the Japanese invasion. In September an ambulance unit was also dispatched to Canton, which was a victim of Japanese bombardment. All this symbolized, as Bose stated, 'India's goodwill, esteem and sympathy for the Chinese nation'. He had no illusion about the behaviour of the Nazis in Europe. With specific reference to Munich Pact, which he called as 'an abject surrender to Nazi Germany'[27], he wondered how 'Europe had sold her soul for the sake of a seven day's earthly existence' and passed the hegemony in Europe into the hands of Germany! [28]

It may nevertheless be mentioned that, while Bose was an uncompromising player in the game of politics, he was also accommodative of the views of others in the Congress who had advocated a policy of active opposition to both fascism and imperialism everywhere. He 'understood the strong sentiment in the Congress behind it and though not approving, not only did not protest but publicly associated himself with it as president'.[29] But Bose was cold-shouldered by the Congress hierarchy, which had adopted manifestly an anti-Fascist approach, being strongly influenced by the views of Gandhi and Nehru in its foreign policy orientation, sought to negotiate with Britain, denouncing the Axis alliance. Not only he was isolated in the Congress for his views, forcing him to resign but was also expelled unceremoniously, and the Bengal Pradesh Committee which had backed him up was also banned and immobilised in an intolerant fashion. It was in sheer disgust and remorse he stated in Delhi on 12 October 1939 that he was 'opposed to Hitlerism whether in India, within the Congress or any other country'.[30]

Bose was not particularly prejudiced about Britain except for the question of British imperialism and colonisation of India. He certainly cherished no 'enmity', as he said, 'towards the British people. We are fighting Great Britain and we want the fullest liberty to determine our future relations with her. But once we have real self-determination, there is no reason why we should not enter into the most cordial relations with the British people'.[31] It seems obvious that this attitude towards the Axis alliance was merely tactical and very much consistent with classical strategic theorising: 'the enemies of Britain, he thought, in the context of the War, 'were the friends of India and the defeat of Britain in the War would lead to emancipation of India' or to put in more categorical terms in his own words, The fundamental principles of our foreign policy has been and will be: Britain's enemy is India's friend".[32]

Long before the outbreak of the Second World War, Bose was convinced of an impending international crisis involving the major powers and he wanted that India should make the fullest use of that crisis to win back her freedom. Naturally he considered the war of 1939 as 'a blessing in disguise', which had set the final stage of struggle for India's liberation. Having witnessed 'an unprecedented awakening in the Indian states' and with his firm belief that India was ripe for revolution, he perceived 'as a cold-blooded realist' that an 'opportune moment' had arrived in India's national history for a final showdown with the British when everyone seemed to have been willing to give blood necessary so as to win liberty.[33] All that it needed then to sink internal 'differences, pull our resources and pull our weight in the national struggle' and seek international support and one all-embracing organisation to link up Indian nationalists all over the world. Therefore, he pleaded that in their own national interest the Indians should dissociate themselves from the British war efforts, as any British victory in the war is likely to delay India's independence.

Initially Bose was not interested in mobilising support of the Axis alliance, although he did understand, for instance, "the imperialistic logic, which worked in Tokyo 'on her race for expansion' and 'act of aggression' which took the world opinion by surprise."[34] He contemplated going to China, but the British authorities refused to permit him to visit China and even threatened to intern him under the Defence of India Act.[35] Having a view of the USSR as an anti-British and anti-imperialist power he had a desire to go to Russia, as he felt that 'Russia is the only country which can help to liberate India'. Moreover, the attack from the North-West was in keeping with traditional historical precedent, but the declaration of the Russo-German War dashed his hope for inducing the Soviet support from the North-West towards Indian liberation struggle at that time. Therefore, he had to re-plan his international strategy, with the planned assault from the non-traditional direction, the North-East, a strategy which 'in its originality, planning and daring is unequalled in the world's military annals".[36]

It may be mentioned that Bose's attempt to induct support of the Axis powers was no more than a response to Britain's own war effort, as it had introduced troops and weapons from other foreign countries in India. It was only natural that the nationalists fighting the British would have the similar right to seek foreign support against the British imperialist intrusion. But, he insisted, 'the emancipation of India must be the work primarily of Indians themselves'.[37] As far as Britain was concerned he did not conceal his rules of the game. He was a zero sum player and therefore, bluntly stated: 'I am an extremist and my principle is all or none'.[38] He would sacrifice everything so as to attain the objective of India's independence.

The INA itself certainly had failed in its immediate mission in Arakan or in the Imphal Operation, but the failure itself may be more ascribed to an overextended Japan (which had deferred the planned offensive against the enemy until the monsoon of 1944) than to the lapses of the INA or of its Commander-in-Chief. After the failure in this front, Bose, as Commander-in-Chief of the Azad Hind Fouz had to leave Burma, as he said, 'with a heavy heart; yet with an invincible spirit he implored others not to be disheartened, as India, he felt, might have 'lost the first round of our fight for independence. But ... there are many more rounds to fight'.[39] For, as he said, 'the roads to Delhi are many, like the roads to Rome. And along one of these many roads we shall travel and ultimately reach our destination, the metropolis of India'.[40]

Even the defeat and surrender of Japan did not disillusion him, as he asserted that 'the darkest hour precedes the dawn'. That 'Japan's surrender was not India's surrender' and that the justness of India's liberation was such that it was 'bound to prevail in the long run'.[41] With his hindsight of shifting diplomatic and strategic dynamics he was certain that the alliance between Russia and the West would—not outlast the war in Europe and that USSR would prove to be a greater menace to Anglo American imperialism than the Axis powers were. Indeed, being aware of Soviet sympathy for a free India, Bose's Provisional Government of Azad Hind had never declared war on Russia, which Bose had always considered as an anti-imperialist power, and 'not an enemy' in so far as India's independence was concerned, although the Provisional Government did declare war on both Britain and the USA. Hence after Japan's surrender to the Allies he wished to go to Russia to seek Soviet help to fight the British. Following his insistence it was indeed arranged for Bose to fly by a special aeroplane to Manchuria, then under Russian occupation, but the ill fated plane crashed in Taipei on 18 August 1945 on its way to Dairen, resulting in his untimely death as well as a sad demise of his scheme of inducting Soviet support for India's liberation.

Bose, indeed, was a nationalist, not a stooge of any foreign power. He had full knowledge of world politics and a masterly understanding of changing international situations. His consistent emphasis on a strong central government for a liberated India was based not only on the requirements of internal consolidation but also on the considerations of ensuring security against any external attack. He had the boldness and courage to protest the disparaging comments made by Hitler in his *Mein Kampf* on the ability of the Indians to 'to rebel against the British', which also attested to his dislike of Hitler and his Nazism.[42] The government he headed and the Army he commanded maintained their independent postures, even during the worst of crises during the war. The INA did, of course, work out a common strategy with the Japanese against the perceived common enemy, but it did operate quite independently, that is, without playing any role unless the common strategy so demanded.

Thus, in 1944 while the INA was enabled to participate in the Imphal Operation as a full and equal participant, the Japanese were refused permission to launch their planned heavy air attack on Calcutta as it would alienate Indian sentiment and cause unnecessary suffering and panic among the Indian people. While formal recognition from the powers allied with the Axis powers in October 1943 did provide the Provisional Government of Azad Hind with international legitimacy, Bose's Government did not necessarily toe the Axis line on war and peace nor did Bose give away anything, not even symbols, which might have harmed India's interests and honour.

Rather, Bose persuaded the Japanese government to transfer the Japanese held Indian territory of Andaman and Nicobar Islands in the Indian Ocean to the Provisional Government of Azad Hind, renamed respectively as *Shaheed* (Martyr) and *Swaraj* (Freedom), so that he could assert his sovereign authority, even if for propaganda purpose. He maintained the diplomatic dignity and honour of India even after the debacle in Burma, as he refused to receive the Japanese Ambassador to Bose's government, Hachiya, who came to Rangoon without any diplomatic 'credentials' and had to wait until they were produced. In similar vein he refused to allow the INA to be used against the Burmese National Army of Aung San after its revolt against the Japanese. He even refused to accept the honour of 'Order of the Rising Sun' offered by the Japanese Emperor until India was free and was in a position to reciprocate.

Bose's thought on international system was to replace the old hierarchy of order with a new pattern of symmetrical interaction from below involving nations then under colonial subjugation; the idea was to circumvent the asymmetric order of international relations of the time, led by some metropolitan empires. In such a prognosis he seems to have been far advanced of his time. He also advocated internationalism, not of the existing British Commonwealth or of the Japanese sponsored 'Co-prosperity Sphere' type, both of which then emerged as mere imperialist tools.

Bose had laid his faith in regional systems based on common interests and culture, which would gradually extend until a symmetrical order of international system came into being. His view of the then British Commonwealth was that it must develop into an association of free nations or disintegrate. It hardly needs mentioning that the post-war transformation of that body was consistent with Bose's vision. Similarly, he felt that the 'Co-Prosperity Sphere' as a regional framework might help promote world peace minus the phenomenon of Japan's hegemony. In the context of the changing global scenario in the post-Cold War era, especially in the context of developing relationships in the Asian-pacific region where Japan may be seen as a major player, one may perceive the emergence of a new regional grouping in the region, seemingly consistent with the project of Bose.

Subhas Chandra Bose was a pragmatic revolutionary strategist and a politician of top order who had to innovate conceptual ideas on strategy at a time when there was very little intellectual framework for coping with the practical problems of politics. Bose's conceptual ideas were, indeed, unique in the sense that the domestic frame of his multifaceted struggle was inseparable from the external linkages that he sought to develop and cultivate. Thus, his vision of politics in general was an all-encompassing one and in that sense one might suggest that he laid his faith in a general theory of strategy with all its linkage components. Having developed his framework he sought to apply all his theoretical concepts to test them on the slippery ground of politics of his time.

Basically, Bose's struggle was for a state that was non-existent as a free political entity and for a country that had lost its dignity in the comity of nations because of an imposed colonial bondage. His youthful dream, as he wrote to his mother, was merely to be 'an adventurer'[43], but in the end he emerged to many Indian revolutionaries as the 'Man of Destiny' or 'Netaji'—the revered leader—who could have delivered the Indian nation not merely its freedom and independence but also the much cherished development, equality and social justice as well as restore India's lost prestige in the comity of nations; but the fateful events of history had cut short his mission and did not permit him to immediately deliver what he had visualised.

In a broadcast over Azad Hind Radio, Subhas Bose talked about his methodology of national struggle. He said that the campaign that was going on in India was non-violent guerilla warfare. He asserted that the object of this guerilla warfare would be of two fold: one, to destroy war production in India, and the other, to paralyse the British administration. All the people of India should participate in the struggle. He chalked out the detail programme for the people. He suggested that payment of the taxes be stopped which directly or indirectly brought revenue to the government. Workers in all industries should either launch a strike or try to hamper production by resorting to a go-slow campaign inside the factories. They should also carry out sabotage with methods such as the removing of nuts and bolts in order to impede production. The students should organise secret guerilla bands for carrying on sabotage activities in different parts of the country. They should also invent new ways of annoying the British authorities, for example, burning stamps, etc. in post offices, destroying British monuments, etc. Women, especially girl students, should do underground work of all kinds especially as secret messengers or to provide shelter for the men who are fighting the imperialists. The government officials who are prepared to help the campaign should not resign their posts but those in government offices and in war industries should give all available information to fighters outside and should try to hamper production by working inefficiently. The servants who are working in the houses of Britishers, should be organized for the purpose of giving trouble to their masters, for example, by demanding higher salaries, cooking and serving bad food and drinks, etc. The Indians should give up all business with banks, firms, insurance companies, etc.[44]

For general public he suggested the following activities:

(a) total boycott of Britishers in India and those Indians who are pro-British;
(b) boycott of British goods, industry, burning of British stalls and government stores;
(c) demonstration inspite of official prohibition;
(d) publishing of secret bulletins and setting up of a secret radio station;
(e) marching to the houses of British government officials and demanding their departure from India;
(f) organising of processions for entering and occupying government offices, secretarial buildings, law courts, etc. with a view to hampering administration;
(g) arranging to punish police officers and prison officials who oppress persecuted people;
(h) begin erecting barricades in the streets where there is a likelihood of attack from the police and military;
(i) setting fire to government offices and factories which are working for war purposes;
(j) interrupting postal, telegraph and telephone communication, as frequently as possible;

(k) interrupting railways, bus and tram services, whenever there is a possibility of hampering the transport of soldiers or of war material, and

(1) destroying police stations, railway stations and jails in isolated places.[45]

REFERENCES

1. *Calcutta Municipal Gazette*, No. 6, June 20, 1925.
2. *Ibid.*
3. Dr. Manju Chattopadhyaya in *Centenary Volume*, published by the Government of West Bengal, April, 1997.
4. Brailsford, H.N., *Rebel India, op. cit.*
5. Chatterjee, Santimoy, *Collected Work of Meghnath Saha, op. cit.*
6. Bose, Subhas Chandra, *The Indian Struggle 1920-1940, op. cit.*
7. *Ibid.*
8. Dasgupta, H.N., *Subhash Chandra* (Delhi: Government of India, Publications Division), 1963, pp. 312-313.
9. *Ibid.*
10. *Calcutta Municipal Gazette, op. cit.*
11. Kalam, Abul, *Netaji Subhas Bose: Strategic Concepts and Diplomatic Thought* (Delhi:),1997.
12. Toye, Hugh, *The Springing Tiger: Subhash Chandra Bose, op. cit.*, p. 177
13. Bose Sisr K., *Netaji Collected Works, Vols.* 1-8 (Calcutta: Netaji Research Bureau),1980, pp. 306-307.
14. *Bose and Sinha*, 1979, pp. 166-167.
15. Toye, Hug, *The Springing Tiger: Subhash Chandra Bose, op. cit.*, pp. 84-85, 103.
16. Basu, Sankariproposad, "Londone Subhas Chandra" in Biswanath De (ed.), *Subhas Smriti (Reminiscences)* (Calcutta: Sahityam), 1981, p. 111.
17. Chakarvarty S.R., and Madan C. Paul, *Netaji Subhas Chandra Bose Relevance to Contemporary World* (New Delhi: Har-Anand), 2000, p. 30.
18. *Ibid.*, p. 31.
19. Dasgupta, H.N., *Subhas Chandra, op. cit.*, p. 321.
20. Bose Sisir K., *Netaji Collected Works*, Vols. 1-8, *op. cit.*, p. 310.
21. Basu, Aparna, *The Growth of Education and Political Development in India 1898-1920* (Delhi: Oxford University Press), 1980, p. 7.
22. Dasgupta, H.N., *Subhash Chandra, op. cit.*, p. 164.
23. Bose Sisir K., *Netaji Collected Works*, Vols. 1-8, *op. cit.*, p. 16.
24. Dasgupta, H.N., *Subhas Chandra, op. cit.*, pp. 322-324.
25. Toye, Hugh, *The Springing Tiger: Subhash Chandra Bose, op. cit.*, p. 70.
26. *Ibid.*, pp. 118-119.

27. Dasgupta, H.N., *Subhas Chandra, op. cit.*, p. 164.
28. *Ibid.*, p. 356.
29. Bimal Prasad (ed.), *A Revolutionary Quest* (Delhi: Oxford University Press), 1980, p. 125.
30. Dasgupta, H.N., *Subhas Chandra, op. cit.*, p. 338.
31. Toye, Hug, *The Springing Tiger: Subhash Chandra Bose, op. cit.*, p. 52.
32. Dasgupta, H.N., *Subhash Chandra, op. cit.*, pp. 250, 294.
33. Das, Cittaranjan, *India for Indians* (Madras: S. Ganesh), 1921.
34. *Ibid.*, p. 112.
35. Bose Sisir K., *Netaji Collected Works, Vols.* 1-8, *op. cit.*, p. 187.
36. *Ibid.*, p. 74-75.
37. Dasgupta, H.N., *Subhash Chandra, op. cit.*, pp. 222-223.
38. *Ibid.*
39. Toye, Hug, *The Springing Tiger: Subhash Chandra Bose, op. cit.*, p. 36.
40. Bimal Prasad (ed.), *A Revolutionary Quest, op. cit.*, p. 80.
41. Bose, Subhas Chandra, *Netaji Collected Works*, 12 Vols. (Calcutta: Netaji Research Bureau), 1980 1996.
42. Bose, Subhas Chandra, *The Indian Struggle* (Calcutta: Published for Netaji Publishing Society), 1948.
43. *Ibid.*
44. Chakarvarty S.R., and Madan C. Paul, *Netaji Subhas Chandra Bose: Relevance to Contemporary World* (New Delhi: Har-Anand), 2000, p. 36.
45. *Ibid.*, pp. 36-37.

4

CONCEPTION OF SOCIALISM

In 1931, Subhas did not 'hold to non-violence on principle'and neither did he think that an anti-imperialist insurrection was imminent in the country. He had, by then, perceived the relevance of Ireland's experience to India's freedom's struggle, especially as both countries had in British imperialism the common adversary. No two situations were wholly alike. However Bose was struck by the fact that both in Ireland and India, British imperial policy had fostered religious divisiveness to defeat the demand for independence. In both countries the elective legislatures offered limited opportunities which could be used to strengthen the popular movements conducted outside for independence. He was especially impressed by the ingenuity with which Ireland's Fianna Fall and Sinn Fein parties, in the single-minded pursuit of the goal of independence, had made use of the 1914-18 World War for Irish independence.

Bose's interest in the reformist endeavours in Turkey, Persia and Afghanistan in the 1920s was natural. These were underdeveloped countries in the thorns of modernization. Even Egypt was relevant in his estimate because of the constitutional advance the Wafd Party had extracted from Britain.

By 1931 the militarist wing of Japan's elite had triumphed over the civilians, and the Japanese military had embarked on a vast expansionist enterprise with the immediate conquest of Manchuria. China's travails thus were of direct interest to him. Japan, home to many Indian revolutionaries who had escaped British retribution, alternately attracted and repelled Indian nationalists who scanned the developments in the East from India. But Italy's fascist coup was altogether different. Involved here was not the Indian struggle for independence and unification, but ostensibly an altogether new socio-political experiment. It was different from the Ata Turk model of post-war Turkey in that Mustafa Kemal Pasha's reforms made no claim to their universal applicability. In so far the Turkish experiment was relevant to non-Turkish societies, they offered a direction to similar traditionalist underdeveloped countries.

Whereas Turkey's Kemalist revolution was precipitated by its defeat in the First World War, Italian fascism took birth in a country which was one of the victorious Allied Powers. Normally it should have had little reason for wanting to overturn the international order the Allies had enforced as the victorious side of the First World War. Yet fascist Italy, while enjoying the territorial and colonial gains conceded to it by the post-war treaties and inter-Allies settlements, became a major source of de-stablization in the interwar years. Tsarist Russia ceased to be an Allied Power after the 1917 revolution.[1]

Italian fascism therefore had its roots in reasons other than only dissatisfaction with the Versailles Peace Treaty and the other pacts and settlements reached in the years following the armistice of 1918. It claimed to provide a socio-economic doctrine of development, rejecting the models of liberal "plutocracies" of Europe and North America on one side and Bolshevism at the other end. In so far as Italian fascism attracted attention in India in the 1920s, broadly speaking, the attraction derived from three elements. First, it was undermining the world order imposed by the main colonial powers, Britain, France, Holland and America. Asia's national freedom movements, being suppressed by these colonial powers, shared a common, visceral hatred of these powers and instinctively viewed with favour any power opposing them. This was especially so before Benito Mussolini grabbed parts of Britain's African colonies and invaded Abyssinia. Second, Italian fascisim exalted nationalism. As the new nationalism of the colonies inspired all national liberation movements, especially India's long drawn out freedom struggle, fervent Indian nationalists could empathize with the stress on Italian nationalism, especially because Italian fascism, unlike Germany's Nazism ten years later, was not overtly racist or anti-semitic. Finally,

Italian fascism, in the early years laid heavy stress on order, discipline and a paramilitary youth movement all of which were of paramount interest to a country like India.

The disparate population and disorder caused anxiety to those Indian nationalist leaders who thought of national reconstruction after independence. Comments in the Indian nationalist newspapers and journals initially on the fascist experiment in Italy reflected inquisitiveness. But these were not judgmental at this stage. In 1932 Jawaharlal noted Bolshevism and Fascism as the two upcoming trends in modern Europe.

On 24 September 1930, a day after his release from the Alipur Central Jail in Calcutta, Subhas took charge as the Mayor of Calcutta, to which position he had been elected in August 1930 while he was serving his term in prison. Six years had elapsed since his appointment by Deshbandhu as the Chief Executive Officer of the Calcutta Corporation and his arrest and detention under Regulation III of 1818 in Burma's jails. Of these six years, nearly four were spent in prison, during which period he had felt obliged to take recourse to a hunger strike for fourteen days to secure the right of detenus to celebrate the annual autumn worship of Durga.[2] On 22 April 1930, Subhas was so brutally assaulted by prison warders in the Alipur Central Jail that he fell down the steps, lost consciousness and had to be carried to his cell by his fellow-prisoners. He received no immediate medical attention. He thus experienced, first hand, the brutalities and long imprisonments to which generally the front-rank Congress leaders were not subjected.

The incident occurred four days after the uprising at Chittagong. Some importance attaches to Bose's inaugural speech as Mayor of Calcutta, for it is believed to be the first public avowal by him of the concept of *daridranarayan* as socialism under a spiritual garb. It is again believed to be his first statement (24 September 1930) of his preference for a synthesis of Europe's currently dominant doctrines of socialism (Bolshevism) and fascism. The words he used were cautious though these were to later cause misunderstanding and misgivings.

"I would say that we have in this policy (that is, Socialism in the spiritual garb) and programme a synthesis of what Modern Europe calls Socialism and Fascism".

He then proceeded to explain the aspects which he felt should form part of the socialist policy in India.

"We have here (that is, in the doctrine of socialism C.R. Das would probably have preferred) the justice, the equality, the love, which is the basis of Socialism, and combined with that we have the efficiency and the discipline of Fascism as it stands in Europe today."[3]

The above extract revealed a quest for a new doctrinal basis for the policies required for modernizing Indian society of the future. Although a very tentative statement, it contained two layers of thought. Basic to the proposition was, first, his belief that the teachings of Vivekananda and C.R. Das provided a directive of an egalitarian policy for future India; and that this doctrine furnished the philosophical spiritual nourishing roots of India's socialism. Secondly, neither Marxism-Leninism nor fascism was complete in all aspects, requiring no further development or evolution. This latter train of thought, the claim of completeness by either doctrine, must have struck him as unnatural, and therefore unacceptable. It took him his release from three years' imprisonment in 1927 (at the Maharashtra Political Conference at Pune on 3 May 1928). He stressed again the indigenous inspiration of socialist policies on 30 March 1929 (at the Rangpur Political Conference), in the presidential address at the student's conference at Lahore on 19 October 1929, and on 27 March 1931 when as President of the Karachi Conference of the All India Naujawan Bharat Sabha he called for a socialist republic in India. "To summarize what I have said, I want a Socialist Republic in India", said Subhas Bose at Karachi's Naujawan Bharat Sabha Conference.

On 4 July 1931, as the President of the All India Trade Union Congress session in Calcutta, he spoke on behalf of those who wanted work for "full-blooded socialism", in the process evolving India's own form of socialism, employing her own methods. In his avowals of socialism he never spoke of Karl Marx as the source of his beliefs, although he was not unfamiliar with the corpus of thought described as Marxism, the "full-blooded socialism" that he wanted was never at this stage described as "scientific socialism", a description the communists favoured to distinguish themselves from other protagonists of socialism.

The estimates made of young Indian nationalist leaders on the strength of the public statements they made in a given period cannot always be accurate. This is because the assessments attributed to these young leaders omniscience and wisdom to which they made no claim. Often they were responding to distant developments and persons of whom they had no personal knowledge. Their opinions influenced policies adopted by the premier nationalist front, the Congress, and Indian public opinion generally. But this was a natural consequence of the positions they occupied in the country's public life. This would not justify, seventy or sixty years later, an expectation that these young leaders should have possessed a prophetic insight to know the future of the developments they commented upon. It is not fair to hold them to every statement they made on major world developments, as though they had been authors of India's own autochthonous doctrines, whereas in most cases they were merely reacting to developments whose outcomes were decided not by them, not even in India.

Examining their responses, from this distance, one is struck not so much by the errors they shared with their many contemporaries. More noteworthy is the fact that these statements were insightful and remarkably close to an understanding of the trends underlying these vast developments in foreign lands. For example, Jawaharlal and Subhas in the 1930s took up positions on major developments of the interwar years. On most issues, notably those taking the world towards another war, the unredeemed evils concealed in the apparent successes of a variety of dictatorships in Europe of the interwar years, the long range effects of Japan's imperialist ambitions in Asia and the effects another total war would produce on India's freedom struggle, their positions were similar. They differed, however, in their attitudes to the communist movement of which the global centre was located in Moscow. As both were Congress leaders, wholly pre-occupied with the domestic issues, these differences, until the Second World War changed the situation, affected them only indirectly, notably by the varying responses they evoked from the Communist controlled bodies in India and abroad.

Chattopadhyaya's letter to Nehru branding Subhas as reactionary has been mentioned before. In 1933-34 Jawaharlal made several statements which were more explicitly supportive of the Soviet experiments; and Subhas, then in exile in Europe, took notice of them, especially Nehru's press statement of 18 December 1933, in *The Leader* issue of 20 December 1933.

"I do believe that fundamentally the choice before the world today is one between some form of communism and some form of fascism, and I am all for the former, that is communism. I dislike fascism intensely and I do not think it is anything more than a crude and brutal effort of the present capitalist order to preserve itself at any cost. There is no middle road between fascism and communism. One has to choose between the two and I choose the communist ideal ... I may not agree with everything that the Orthodox communists have done ... But I do think that the basic ideology of communism and its scientific interpretation of history is sound."[4]

Subhas Chandra Bose was a believer in socialism. He asserted that he wanted a 'Socialist Republic of India'. However, his concept of socialism was different from those of others which he called as Indian socialism. Addressing Bhartiya Naujawan Sabha in 1931, he said: "If we undertake a comparative analysis of different social and political ideals that have inspired human endeavour and activity throughout the ages, we shall arrive at certain common principles that should form the basis of our collective life which are justice, equality, freedom, discipline and love. In order to be just and impartial, he advocated to treat all men as equal. He asserted that bondage of any kind, economic or political, robs men of their freedom and this gives rise to inequalities of various kinds. Therefore, in order to ensure equality it is necessary to get rid of bondage of every kind: social, economic and political. However, he also warned that freedom did not mean indiscipline or absence of law. It rather means a combination of law and discipline. This is very necessary for the struggle for freedom which is the basis of life. Among these fundamental principles Bose asserted, love as the highest principle. Without a feeling of love for humanity, one could neither be just to all nor would treat all men as equal, and would neither desire to be called upon to sacrifice. And without sacrifice, freedom could not be of right sort. Thus, for Bose, the cardinal principles of socialism are justice, equality, freedom, discipline and love. He advised people of India not to ignore the history and tradition of the country while adopting the social and political institutions of other countries. He said that there were all kinds of active nation building programmes, social and political ideologies in the Western world, such as socialism, state socialism, guild socialism, syndicalism, philosophical anarchism, Bolshevism, Fascism, parliamentary democracy, aristocracy, absolute monarchy, dictatorship etc. He accepted some wisdom in each of them. However, he warned that in a progressive world like ours it would not be proper to accept any one of them as the last word for an ideal or the final solution of all our social and political problems. Because, he reasoned that the results of the transplantation of an entire idea or institution in a country from another may not be necessarily agreeable or fruitful. A national institution is the natural result of history of the people concerned, their thoughts and ideals, and the activities of their day-to-day life. Therefore, it should be borne in mind, according to Bose, that social and political institutions could not be built by ignoring the history and tradition of the people of the country, their present condition or prevailing atmosphere of life. For the above reason, he could not agree with the Communists in India also.

He felt that Bolshevik socialism of Soviet Union did not suit India. There should be assimilation of only those socialistic principles which could suit Indian requirements. He explained that Bolshevik theory had been passing through experimental stages and the communists were departing from Lenin and other Bolsheviks. This departure had been caused by the peculiar conditions and circumstances prevailing in Russia which have compelled a modification of the original theories. At another place he warned against the unsuitability of Marxian ideas. He said that "Marxian ideas" coming from the West are like 'boisterous breakers', and some people in India are excited and stirred up with these. Most of them, said Bose, believe that the adoption of Marxism in its pure and complete form would make India a land of joy and plenty, and point their fingers towards Russia. In fact, they are wrong because Bolshevikism adopted and practised in Russia had its points of disagreement with Marxian socialism. Because people of Russia did not ignore their ancient traditions and national ideals. He asserted that had Marx been alive, he would not have liked the way his ideals had been modified in Russia.

The other reason for rejection of Communism, according to Bose, was the methodology and tactics the Communists employed to achieve their ends. These methods and tactics, he thought, did not suit India, and that is why Communism could not make much headway in this country.

Thus, Indian socialism would have to be different, he asserted. He said that at that stage it was difficult to chalk out the details of a socialist state of his dreams. He could only give an outline, the main features and principles of the socialist state. He said, "We want political freedom whereby it meant the constitution of an independent Indian state, free from the control of British imperialism. It should be quite clear to everybody that independence means severance from the British empire, and on this point there should be no vagueness or mental reservation. Secondly, we want complete economic emancipation. Elaborating his views on economic emancipation, he held that every human being must have the right to work, and the right to living wage. There shall be no unearned income. There must be equal opportunities for all and there should be a fair, just and equitable distribution of wealth. For this purpose, it may be necessary for the state to take over the control of the means of production and distribution of wealth.

Besides economic equality, Bose also felt that the third essential feature of real socialism is the complete social equality. Social equality means that there shall be no caste, or depressed classes. Every man will have the same rights, the same status in society. Bose was not only a believer of eradication of caste hierarchy but also the supporter of women's equality in social status or in law. That a woman will always be an equal partner of man, was his firm faith.

Social Change

Nevertheless the ideal of socialism could not be achieved without social change. Bose had specific views about social change. He was of the view that social reforms and movement for national freedom are inextricably related. Speaking as President at Hooghly District Students' Conference on July 1929, he said that both, those who imagine that they can make their country politically free, leaving its social order untouched or those who think change is possible without political revolution, are wrong. He explains, the former are wrong because once the desire for freedom was aroused in people it could not be halted to wait for society to change. Similarly, the latter are wrong because the society as 'tattered' as ours and, social repression and economic inequality so dominant that people who suffer could not be inspired for political independence. He questioned that how could those people who felt bitterly the string of social and economic exploitation respond to the call of political freedom unless they were assured of freedom from these exploitations.

Thus, a kind of social revolution had been envisaged by those which he thought was inevitable. He asserted that there was no fundamental difference between evolution and revolution. Revolution is nothing but evolution taking place within a comparatively short span of time, whereas evolution is only another name for revolution, spreading over a longer period of time.

Netaji's socialist credentials are best assessed in terms of a number of vital issues relevant to the post-liberation reconstruction of India. Of many such issues, perhaps the most notable among them was the realization that for a country of India's dimension, beset with staggering poverty and unemployment, there could be no escape from planning.

In its meeting at Wardha held between August 14-17, 1937, the Congress Working Committee had passed a resolution relating to the appointment of an Expert's Committee to explore the possibilities of an All India Industrial Plan. In his Haripura Presidential address, Netaji unveiled a secular, broad based blueprint for national reconstruction. Under his Presidentship in July 1938, the Congress Working Committee, referring to the Wardha resolution of August

1937, resolved that "as a preliminary step the President was authorized to convene a Conference of the Ministers of industries and call for report of the existing industries in different provinces and the needs and possibilities of new ones." Netaji outlined his concept of National Planning in his address delivered at the Industries Ministers' Conference on October 2, 1938. Prior to this, in May 1937, he had convened in Bombay a Conference of the Premiers of seven Congress Provinces. Thus having armed himself with the authority of his party and the concurrence of the Government formed by the Congress in seven provinces, Netaji went ahead with the formation of the National Planning Committee, which he inaugurated in Bombay on December 17, 1938.

Netaji took a deliberate decision to appoint Jawaharlal Nehru as the Chairman of National Planning Committee. Offering Chairmanship to him in a letter dated October 19, 1938, he hoped that he would accept the Chairmanship, as its success depended on him.

The Indian Planning Commission, in its commemorative volume on Netaji states:

"Netaji could have assumed the historic role of being the first Chairman himself. But he knew that for planning to be a success, consensus was a must. The period of history in which he headed the Indian National Congress was marked by sharp ideological contradictions within the national movement. Thus, in order to take everyone along and to ensure that the seeds of planning and industrialization in India were sown on fertile soil, Netaji resisted the arid idea of himself dominating the first National Planning Committee though the formation of the Committee was made possible primarily because of the accelerated efforts undertaken by him as the Congress President."[5]

Before the advent of Netaji era, the Congress Working Committee had only been talking of an industrial plan. Netaji introduced the concepts of poverty and unemployment, national security, and national reconstruction along with the need for rapid industrialization into the concept of National Planning.

Mumbai was chosen as the venue for the first meeting of the National Planning Committee because the erstwhile Government of Bombay Province was to provide the infrastructural support for the functioning of this historic Committee. The Committee was inaugurated by Netaji and presided over by Jawaharlal Nehru.[6]

Focus on Industrialization

Netaji gave due emphasis to the need for the country's industrialization. In an interview with an eminent scientist, Meghnath Saha, he stated that the problem facing India was not limited to industrial recovery, but industrial revolution. In this India could not afford to follow the path of gradualism as in the case of Britain, but one of forced march as in the case of Soviet Russia.

In Netaji's conception, industrialization did not mean "manufacturing umbrella handles and bell metal plates." Elaborating further on the principles of national planning and industrialization, he stated that though from the industrial point of view the world was one unit, we should aim at national autonomy, especially in the field of our principal needs and requirements. This implied the development of 'mother industries', such as power, metal production, machine and tools manufacture, manufacture of essential chemical, transport and communication industries.

The need for skilled and qualified manpower to promote industrialization was foremost in Netaji's thinking. In this context he highlighted the need to tackle the problem of technical education and technical research. He realized the need for our students being sent abroad so that on their return they could help establish the new industries. Regarding technical research, he emphasized the need for setting up of a permanent National Research Council.

It might appear to be somewhat inconsistent that Netaji felt that technical research should be free from government control of every kind. He stated thus:

"It is only in this unfortunate country that government servants are entrusted with scientific research on receipt of princely salaries and we know very well what results have been obtained therefrom."

As a preliminary step towards national planning, Netaji was convinced of the need for an economic survey of present industrial position with a view to securing the necessary data for the National Planning Commission.

Stating the imperatives of industrialization, Netaji stated that this was a must to fight the menace of poverty, to provide a scientific base for agricultural transformation, to provide a basis for socialism, and to enable India to be competitive in the world market.[7] Netaji considered the need for setting up of a modern national army as one of the country's topmost priorities at the dawn of the country's independence. For this also there was no escape from industrialization to match the country's requirements for armaments and defense equipment. At a speech delivered at Tokyo University in November 1944, he stated:

"We do not believe that India can achieve freedom without the use of arms. Now once you take up this attitude, that for winning freedom we have to fight and to use arms, it follows that we must have modern industries to manufacture the arms. So we take our stand on modernism. We have to fight the enemy with modern methods and with modern arms, so we must have modern industries".[8]

Netaji was convinced that the staggering problem of poverty and unemployment in India could not be resolved without due emphasis to industrialization. While acknowledging the need for improving the efficiency of agriculture by adopting scientific methods, he was convinced that this would yet fail to solve the problem of poverty and unemployment. In fact, scientific agriculture would result in more being produced by less number of peasants which could in fact, exacerbate the existing problem of unemployment in the rural sector. He felt that there was no option but to transfer a large proportion of rural population to the industrial sector.[9] Netaji's thinking in this field reminds one of Rangnar Nurkse's and, Fei and Ranis's concept of balanced agricultural industrial transformation.

Netaji tried to dispel the view being projected by some Congress leaders that he was out to marginalise the significance and importance of small scale village and Khadi industries. In his inaugural speech at the first meeting of All India National Manning Committee at Bombay on December 17,1938, Netaji stated that there was no inherent conflict between cottage industries, and large-scale industries. Further elaborating, he stated that in fact, the industries needed to be classified in three classes: cottage, medium scale, and large scale industries. He felt the need to work out a plan that could lay down the scope for each of these classes. Among large scale industries, "mother" industries were the most important as they aim at producing the means of production. They put in the hands of artisans improved implements, which could improve their productivity and thereby help them to improve their standard of living. He felt that if basic industries such as power and machinery manufacturing industries were controlled by the state for the welfare of the nation, light industries such as manufacture of bicycles, fountain pens and toys etc. could be started by the artisan class working with the family as the unit. Only when raw materials and basic inputs were provided to them at reasonable price, the handloom and cottage industries could be revived. The aforementioned principles clearly reveal that Netaji was aware of the inter-industry linkages that characterize what Hershman had theorized, the strategy for industrial development. Netaji clearly recognized the need for cheap inputs from basic industries as an essential prerequisite for small scale industries to prosper and to move to higher levels of technical efficiency. His concern for rational classification of industries remains valid even today.

Centralization and Regionai Autonomy

In his Haripura address, Netaji drew attention to the threefold objectives of national reconstruction: (i) to prepare the country for self-sacrifice, (ii) to unify India, and (iii) to give scope for local and cultural autonomy. The last two objectives, he explained, were not contradictory. He pointed out that while unifying India through a Strong central government, there was a need to put all the minority communities as well as the provinces at their ease, by allowing them a large measure of autonomy in cultural as well as governmental affairs.

In an article published in a German periodical, Netaji tried to further elaborate on the question of reconciling the need for a strong Central Government with that of regional autonomy in these words:

"One thing, however, is clear. There will be a strong Central Government. Without such a government, order and public security cannot be safeguarded. Behind this government will stand a well organized, disciplined all India party, which will be the chief instrument for maintaining national unity.

The state will guarantee complete religious and cultural freedom and there will be no state religion. In the matter of political and economic rights there will be perfect equality among the whole population. When every individual has employment, food and education, and has freedom in religious and cultural matters, there will be no more any minority's problem in India.

When the new regime is stabilized and the state machinery begins to function smoothly, power will be decentralized and the provincial governments will be given more responsibility". It thus appears that Netaji visualized a strong central government as an interim measure only after which more powers could be transferred to the provinces.

When the new regime is stabilized and the state-machinery begins to function smoothly, power will be decentralized and the provincial governments will be given more responsibility."[10]

It thus appears that Netaji visualized a strong central government as an interim measure only after which more powers could be transferred to the provinces.

Socialistic Pattern of Society: Twin Principles of Nationalism and Socialism

Bose's idea of a socialistic pattern of society was developed chronologically and logically over the years. The main points of thoughts were:

(1) There needs to be a planned programme to revive and develop the agriculture and industry of the nation after independence.

(2) Panchayat System of Government will have to be introduced in the villages in order to introduce political self-government, which will eradicate the stifling age old customs and practices like the caste system.

(3) Zamindari system will have to be abolished and a modern system of fiscal management like loans given by banks to be introduced through banking and other financial institutions.

(4) Production units will have to be modernised and distribution made equitable.

"Free India will not be a land of capitalists, landlords and castes. Free India will be a social and political democracy," he said. Bose argued in favour of social ownership and control of both production and distribution, and setting up of large scale industries for economic regeneration of the country.

Bose's political thought was thus based on twin principles of nationalism and socialism, but it will be erroneous to conclude that he was in any manner a supporter of national socialist thought as of Germany. Outspoken as he was, sometimes his words and writings gave rise to misconceptions and misunderstandings. He repeatedly said that India had to evolve her own idea and form of socialism and her own ways to achieve the goal. Socialism for India will not only be a political idealism, but it will provide a programme for future reorganisation. He was certain that in future, India had a contribution to make on providing to the world a new ideology. In an address from East Asia, given on 21 October 1943, Netaji Subhas Bose gave this assurance that the INA was being prepared not only to wage war against the enemy but for social and economic reconstruction of India after the war. Netaji's speech at the University of Tokyo in November 1944 can be quoted to be perhaps the most fundamental of Bose's ideas in clarifying his ideas and policy. He identified poverty and illiteracy as the two evils that was thwarting progress. He said that eradication of these, and that of unemployment will have to be dealt with through state initiative; for if left to private hands, it would take centuries. Such reforms have to be speeded up and done on a warfooting.

Earlier, in the famous "Kabul Thesis" written after he left home in 1941, he enunciated his justification for organising the Forward Bloc and gave a clear direction to the party, as to what were to be attained after independence:

(1) A thoroughly modern and socialist state;
(2) Scientific large scale production for the economic regeneration of the country;
(3) Social ownership and control of both production and distribution; and
(4) Application of the principles of equality and social justice, in building up of the new order in Free India.

Socialism to Subhas Chandra Bose was as much a political idealism as it was the way to create and organise future India. He gave a complete work plan for the socialist reconstruction of free India. His programme would ultimately create a free and classless society in India. "The salvation of India lies in socialism", he said, but "India should be able to evolve her own methods in keeping with her own needs and her own environment. This socialist system will be developed in an Indian way and under Indian conditions". To quote once again his Kabul Thesis, wherein he had said, "In the present political phase of Indian life, Leftism means anti imperialism. After the attainment of political independence Leftism will mean Socialism and the task before the people will then be the reconstruction of national life on a Socialist basis".[11]

REFERENCES

1. Carr, E.H., *International Relations Between the Two World Wars* (London: MacMillan & Co. Ltd.), 1952, pp. 66-72.
2. "In Durga we Mother, Motherland and the Universe all in one. She is at once Mother, Motherland and the Universal Spirit"—Subhas Bose's Correspondence—his letter from Mandalay Jail to his sister-in-law, Bivabati Bose, wife of Sarat Bose.
3. Subhas Bose's Speech reported in Calcutta Municipal Gazette, 27 September 1930.

4. *Selected Works of Jawaharlal Nehru* (Delhi: Orient Longman), 1974, p. 134.
5. Planning Commission, *Subhas Chandra Bose: Pioneer of Indian Planning* (New Delhi: Planning Commission, Government of India), 1997.
6. Other members present included : Dr. M Viveswaraya, Purusottamdas Thakurdas, Ambalal Sarabai, A.D. Shroff, K.T. Shah (Secretary of the Committee), Dr. V.S. Dubey, and V.V. Giri. The then Secretary of the All India Congress Committee, Acharaya J. B. Kripalani, was present as a special invitee along with M.L. Patil, (Local Self-government Minister, Bombay Province), C.J. Bharuka (Industry Minister, Madhya Bharat), M.K. Pagar, (Industry Minister, Baroda), M.S. Sriramachandra Rao (Industry Minister, Mysore), Vallabhbhai Patel, Bhulabhai Desai, Jamnalal Bajaj, P.P. Advani and Professor C.N. Vakil.
7. Answers to Questions posed by Meghnath Saha to Netaji at the *Third General Meeting of the Indian News Science Association* on 21 August, 1938.
8. "The Fundamental Problems of India": Speech Delivered in Tokyo University, 1944.
9. "The Industrial Problems of India", Address Delivered at the Industries Ministers Conference in Delhi, October 2, 1938.
10. Bose, Chandra Subhas, *The Indian Struggle, op. cit.*
11. Chakarvarty S.R., and Madan C. Paul, *Netaji Subhas Chandra Bose Relevance to Contemporary World, op. cit.*, pp. 44-45.

5

COMMUNAL AMITY AND HARMONY

Netaji Subhas Chandra Bose, an ardent student of Indian history and a spiritual being, considered spiritual religion (not dogmatic religion) strictly a matter of private belief and pursuit. The eternal ideal of *Sarvadharma Samabhava* and the policy of "to live and let others live" shaped the broad outlook and attitude of Subhas Chandra Bose from his early childhood. He honestly endeavoured to establish communal amity among all the people irrespective of caste, creed and religious identities. He followed the footstep of his mentor, C.R. Das, who imbibed the true spirit of his Guru Purushottam Thakur Anukul Chandra Chakraborty. Sri Sri Thakur believed and worked for Hindu-Muslim amity to further the unity of whole humanity. Both C.R. Das and Subbas Chandra Bose deeply believed that the Hindu-Muslim unity was crucial for achieving the goal of India's freedom. Moreover, Muslims constituted a majority of the population of undivided Bengal, that is, about 20 per cent of India's population. Both of them urged the Congress leaders not to ignore this substantial segment of Indian populace. Subhas Chandra Bose despised everything narrow and selfish that came in the way of communal harmony.[1]

Immensely disturbed by the rising communal divide in Bengal and other places in India, and having been convinced of the need of Muslim support, if the Swarajyists were to succeed in blocking the reactionary colonial government rule and policies, C.R. Das worked out a provisional agreement known as Bengal Pact[2] with several Bengali Muslim leaders in 1923. In this, Das promised to provide 60 per cent of all the new appointments in government departments to

Muslims if the Swarajyists were voted to power. He assured similar share of appointments to Muslims if they voted the Swarajyists in Calcutta Corporation election. The Bengal Pact in fact covered both the religious and political questions as well. Das believed this would help strengthen the communal relationships between the two great communities, and thereby, solve the question of communal tension in Bengal. This was also the idea of his Guru Thakur Anukul Chandra Chakraborty of Satsang (now Deoghar).

Das wanted the Bengal Pact to be a part of national agenda of the Congress. And he wanted the Congress party to put a seal of approval on it. But unfortunately, it did not get the approval from the Annual Conference of the Congress which was held in Coconada in Madras Presidency in December 1923. But C.R. Das did not lose heart and remained steadfast in his attempt and weathered all his critics. Unperturbed and untiring, he placed the same Pact again for adoption and approval in the next Annual Conference of Bengal Congressmen (known as Bengal Provincial Conference) held in June 1924 at Serajgunj (North Bengal). Inspite of strong opposition by his political opponents, Deshbandhu Das succeeded in getting the Bengal pact passed by overwhelming majority. By this he showed to the countrymen his strong commitment towards Hindu Muslim Unity. The Muslims of India, especially the Muslims of Bengal, was convinced of his non-communal spirit to establish a solid foundation of communal amity so that the two great communities of India could bind and fight together to attain Swaraj.[3] When the Bengal Pact was passed, Subhas Chandra Bose was the Secretary of the Bengal Provincial Congress Committee (BPCC). In fact the Bengal Pact was published under his signature and seal.[4] Both of them truly believed and worked hard for the Hindu-Muslim unity to throw the British Imperialist out of Indian soil.

As a true follower, Bose treaded the path of his mentor, C.R. Das, the Mayor of Calcutta, and implemented the Bengal Pact sincerely as Chief Executive Officer of Calcutta Corporation. He offered many jobs to the Muslims,[5] even though there was stiff resistance from the reactionary members of Calcutta Corporation. He forcefully countered them and convincingly replied to those who opposed his actions. He said, "I entirely repudiate the suggestion that the appointments have been made in utter disregard of the best interests of the corporation. I am fully conscious of the responsibility that rests on me. I have no Hindu-Muslim Pact before me, but I have certainly taken cognizance of what I consider to be the just claims of the Muhammedans, and in future propose to do so... I am further of the opinion that claims of the different sections of the people have got to be considered in making appointments. I do not think I shall be wrong if I say that in the past Hindus have enjoyed what may be regarded as a sort of monopoly in the matters of appointments... The new Corporation is standing... for a new epoch... has got to respond to the spirit of the times... though it is sure to give rise to a certain amount of heart burning in the ranks of the Hindu candidates".[6]

Netaji Subhas Chandra Bose was unruffled in his stand as Deshbandhu C.R. Das stood firm in his defence alongwith Gandhiji.[7] Subhas was of the opinion that the Congress, including the Swarajyists had to shun communal line of politics to reach out to wider social groups if they want to involve all the Indians into the vortex of nationalist movement. For him Deshbandhu C.R. Das was a real crusader of Hindu-Muslim unity.[8] However, the sudden death of Deshbandhu caused a great tremor to Subhas Bose. In a letter to Hemendra Nath Dasgupta from Mandalay Jail on 20th February, 1926 he paid his highest tribute to Deshbandhu C.R. Das. He wrote: "I do not think that among the Hindu leaders of India, Islam had a greater friend than in the Deshbandhu... Hinduism was extremely dear to his heart; he could even lay down his life for his religion, but at the same time he was absolutely free from dogmatism of any kind. That explains how it was possible for him to love Islam. Let me ask, how many of our Hindu leaders can declare on oath that they do not hate the Muslims? On the other hand, how many of the Muslim leaders can likewise say that they do not hate the Hindus? In religious belief, the Deshbandhu was a Vaishnavaite; but people of all faiths had a place in his large heart. We might settle our disputes through pacts but he did not believe that was the only means of securing cordial relations between the Hindu and Muslim. It was his intention, therefore, to bring about a lasting Hindu-Muslim entente through a kind of cultural synthesis ..."[9]

Perception of Communal Amity

Netaji Subhas Chandra Bose's perception of communal amity and his belief in the separation of nationalist politics from dogmatic religion can be ascertained from several of his speeches, articles and letters written to various persons. In his Presidential address at the Maharashtra Political Conference held at Poona in May 3, 1928 Bose said, "Regarding our communal problem, while I do not condemn any patch up work that may be necessary for healing communal scores, I would urge the necessity of discovering a deeper remedy. It is necessary for the different religious groups to be acquainted with the traditions, ideals and history of one another, because cultural intimacy will pave the way towards communal peace and harmony. The fundamental basis of political unity between different communities lies in cultural rapprochement, and as things stand today, the different communities inhabiting India are too exclusive. In order to facilitate cultural rapprochement, a dose of secular and scientific training is necessary. Fanaticism is the greatest thorn in the path of cultural intimacy, and there is no better remedy for fanaticism than secular and scientific education. This sort of education is useful for another reason. It helps to rouse out "economic" consciousness. The dawn of "economic" consciousness spells the death of fanaticism. There is much more in common between a Hindu peasant and a Muslim peasant than between a Muslim peasant and a Muslim Zamindar. The masses have only got to be educated wherein their economic interests lie, and once they understand that, they will no longer consent to be pawns in communal feuds. By working from the cultural, educational and economic side, we can gradually undermine fanaticism and thereby render possible the growth of healthy nationalism in this country."[10]

In another Presidential address at Rangpur Political Conference on March 30, 1929, Subhas Chandra Bose made certain observations on the true historical tradition of communal harmony among the Hindus and Muslims of Bengal. He explained: "Truth alone is our ideal and that explains why inspite of many inroads on Bengal culture, civilization, literature, and religion, she always assimilated the truth of the newcomers, keeping her individuality intact all the while... As in religion so in literature, Bengal has manifested herself in diverse ways. Her Vidyapati and Chandidas, Mukundaram and Bharat Chandra, Kasiram, Kritibas and Ram Prasad stand as monumental examples of her new discoveries in the domain of thought and culture. Bengal does not forget the contribution made to her literature by the Mussalmans, and it was here that the indissoluble unity between the two communities was formed which has weathered many storms in the past. In a word, Bengal as she is today is the child of universalism, irrespective of any caste and creed... The renaissance of religion was reflected in her literature, philosophy and in all spheres of life, and Hindus and Mussalmans united to preach this gospel of brotherhood."[11]

While reporting to the press on 18 November 1937 about the prevailing communal situation in Bengal, Subhas Chandra Bose pointed out that "today certain sections of the Bengali community are permeated with the spirit of communalism. Out of this poisonous atmosphere the Congress alone can rescue our people, because Congress is the only national organisation in this land. I earnestly hope and trust that as we push on with our economic programme, which shall cut across all communal divisions and unite all sections of our community, I appeal to all those Muslims and members of the scheduled castes who have so far held aloof from the Congress to join the Congress fold. It is for them to capture the Congress and make it their own organisation. If they do so we shall be content to be backbenchers".[12]

In his unfinished auto-biographical work,[13] Subhas Chandra Bose refused to acknowledge the communal periodization of Indian history into "Muslim period', 'Hindu period' etc. For him the Battle of Plassey was a glaring example of Hindu-Muslim cooperation against the common enemy, the British. He proved that Indian history had instances of such Hindu-Muslim unity. Bose tried to dispel the idea generally nursed by many that for a long time under British rule, the Muslims did not play any important role in Indian history, and several theories have been advanced biasedly to account for this. He urged people to understand that since in provinces like Bengal, the rulers who were overthrown by the British were the Muslims by religion, the Muslim community maintained for a long time an attitude of sullen animosity and non-cooperation towards the new rulers, their culture and their administration. On the other hand, it is said that, prior to the establishment of British rule in India, the Muslim aristocracy had already grown rich and that Islam did not at first take kindly to modern science and civilization.

That is why, Bose pointed out that considering the proportion of Muslims in India, they can never cease to play an important role in India. Subhas Chandra Bose candidly said that whether before or under the British rule the distinction between the Hindus and the Muslims of which we hear so much now-a-days was largely an artificial creation of the British Colonialists. It was a kind of catholic protestant controversy in Ireland, in which the present day rulers, the British colonialist, have had a hand. That is why Subhas Chandra Bose rightly reiterated that it is "a misnomer to talk of Muslim rule when describing the political order in India prior to the advent of the British. Whether we talk of the Mughal Emperors at Delhi, or of the Muslim kings of Bengal, we shall find that in either case the administration was run by Hindus and Muslims together, many of the prominent Cabinet Ministers and Generals being Hindus. Further, the consolidation of the Mughal Empire in India was effected with the help of Hindu Commanders-in-chief. The Commander-in-chief of Nawab Siraj-ud-daulah, whom the British fought at Plassey in 1757 and defeated, was a Hindu, and the rebellion of 1857 against the British, in which Hindus and Muslims were found side by side, was fought under the flag of a Muslim, Bahadur Shah".[14]

When the Indian National Army (INA) was formed by Netaji Subhas Chandra Bose he saw to it that it was absolutely free from communal leanings and feelings. He banished all social barriers of caste, creed and religious distinctions in INA both in recruitment and training. He was quite clear in his approach. For him there were no religious or provincial differences as Mother India is one and we are all her children. So, he refused to recognize these differences and trained the INA soldiers uniformly. He had given full freedom to privately practise one's religion in whatever way one might like. He made his soldiers realize that the higher goal of life was unity and thus unitedly liberate Mother India from the clutches of imperial power. In the writings of Major General Shahnawaz

Khan and Major Abid Hussain of INA, the two very close lieutenants and faithful followers of Netaji Subhas Chandra Bose, we find vivid account of non-communal and secular character of the INA and Netaji Subhas Chandra Bose's very careful and sincere endeavour to maintain that character to achieve higher goal of unity to win freedom. Major-General Khan mentioned that for Netaji Subhas Chandra Bose there were no religious or provincial differences. He looked at every one without distinction whether Hindu, Muslim or Sikh and his true secular spirit always animated his followers. There was no "communal" feeling of any sort in INA inspite of the fact that every one had full freedom to practise one's own religion in any way one liked.[15] He proclaimed after the formation of Provisional Government of Azad Hind at Singapore that "the Provisional Government is entitled to and hereby claims, the allegiance of every Indian. It guarantees religious liberty as well as equal rights and equal opportunities to all its citizens."

Netaji made his soldiers realize that their only identity was that they were the sons or daughters of Mother India and as such there could be no differences between them. The soldiers were completely united and they were made to realise that the communal differences in India were the artificial creation of the imperialist British. The success of this experiment in INA can be gauged from the fact that most ardent supporters and admirers of Netaji Subhas Chandra Bose were to be found among the Muslims. Netaji encouraged, inspired and respected every man and woman for what he/she was worth for and not for his/her religion or the province he/she hailed from. It was on this score the *Azad Hind Fauz* refused to believe that it was not possible to unite and live together like brothers and sisters, and work for the creation of a great Indian nation, free and united.[16]

Netaji said that we are eager to do our best to arrive at an agreed solutions consistent with the fundamental principles of *Samyavada* and nationalism. Only by emphasizing our common interests, economic and political, that we could cut across communal divisions and dissensions. A policy of "to live and let others live" in matters of religions and an understanding in matters of economic and political should be our prime objective.[17] While broadcasting from Berlin Radio, Netaji Subhas Chandra Bose assured the fellow countrymen by extolling that he wanted to have a free India with a social order based on the eternal principles of justice, equality and fraternity.[18] So, in his heart of hearts he could never imagine India divided on communal lines. The humanitarian and egalitarian aspects of socialism had great appeal to him. He sought to bring together the two communal parties, the Hindu Mahasabha and the Muslim League, in a joint fight along with the Congress for India's independence. He even suggested that, if necessary, Mohammad Ali Jinnah could become the first Prime Minister of Free India. That is why Netaji assured the Muslims that they had nothing to fear, on the contrary, they had everything to gain in the event of India attaining freedom, as a part of his doctrine of *Samyavada* and secular approach.[19]

To conclude, a deeply spiritual person, Netaji Subhas Chandra Bose was not only a symbol of communal amity but also an uncompromising hero of freedom movement and valliant fighter of British imperialism. He is no more, but his eternal spirit and unorthodox principles are still there to inspire the younger generation of today's India. Netaji's views on communal amity and his sincere endeavours in establishing a solid foundation of non-communal and united India by delinking dogmatic religion completely from society and politics are much more relevant today and should be sincerely followed if his dream is our prime concern. It is then, and only then, his great immortal soul can bless this Motherland for whom he had shed every single drop of precious blood till the last breath of his life.

REFERENCES

1. Saggi, P.D. (ed.), *Life and Works of Netaji Subhas Chandra Bose* (Bombay: Overseas Publishing House), n.d., p. 32.
2. Maulana Abul Kalam Azad, *India Wins Freedom* (New Delhi: Orient Longman), 1989, pp. 23-24.
3. Bose, Subhas Chandra, *The Mission of Life* (Calcutta: Thackar, Spink & Co.), 1965, pp. 56-57.

4. Dutt, Satyabrata, "Subhas Chandra O Muslim Prashna" (in Bengali), *Shatabarshe Subhas Chandra, Parichaya* (Subhas Chandra Special Number), Vol. 65, Nos. 11-12, June-July 1996, Calcutta, pp. 81-82.
5. Gordon, Leonard A., *Brothers Against the Raj* (Delhi: Viking), 1990, p. 112.
6. *Ibid.*, p. 113.
7. *Young India*, July, 1924.
8. Bose, Sisir Kumar., (ed.), *Netaji's Collected Work*, Vol. 5, *op. cit.*, pp. 19-37.
9. *Ibid.*, p. 27.
10. Ayer, S.A., *Selected Speeches of Subhas Chandra Bose* (New Delhi: Publications Division, Ministry of Information and Broadcasting, Government of India), September 1974, p. 33.
11. *Ibid.*, p. 46.
12. Bose, Sisir Kumar., (ed.), *Netaji's Collected Work*, Vol. 1, *op. cit.*, pp. 434-35.
13. An Indian Pilgrim, written in Austria in 1937 which he could not complete.
14. Bose, Sisir Kumar., (ed.), *Netaji's Collected Work*, Vol. 1, *op. cit.*, pp. 14-15.
15. Khan, Shahnawaz, *My Memories of INA and its Netaji* (Delhi: Rajkamal Publication), 1946, pp. 10-11.
16. *Ibid.*
17. Saggi, P.D. (ed.), *Life and Works of Netaji Subhas Chandra Bose, op. cit.*, pp. 15-16.
18. *Ibid.*
19. Bose, Sisir Kumar., (ed.), *Netaji's Collected Work*, Vol. 1, *op. cit.*

6

ON FASCISM

The first notable public reference by Subhas to Fascism hyphenated with socialism was made in his inaugural speech as the Mayor of Calcutta on 24th September 1930. That statement is explicable by the similarities of the single party Statist systems which flourished and commanded widespread popular loyalty in the Europe of interwar years. On these similarities among others of the Statist systems of the age was erected eventually the theory of managerial revolution in the industrially advanced countries.[1]

Apart from those Indian leaders who were guided or influenced by the Communist International's interpretations of world developments, many eminent Indians were slow to awake to the hideousness of the fascist state hidden in Mussolini's rhetoric and the efforts the Duce made to arouse hope of understanding and support in the Indian freedom struggle. Even Europe's awareness of the menace of new dictatorships was not instant. In 1936, four years after the fascist took over in Italy, Romain Rolland sadly remarked in his diary on the prevailing confusion among the otherwise knowledgeable and devout Christians over fascism. Even the truly religious people did not see the incongruity in having Mussolini associated with the centenary celebrations in honour of St. Francis of Assissi! And the Roman Church, far from protesting, welcomed this offensive absurdity.[2]

As Rolland saw it, the suppression of human consciousness by the despotic state in Italy had to be allayed by moral help and advice. Rolland thought of asking C.F. Andrews (1871-1940), a British missionary and a close friend of both Rabindranath Tagore and Gandhi, to come and study the European problems and place his conclusions before the latter to seek Gandhi's advice. Yet Vijaya Lakshmi Pandit visited Italy in 1927 and was struck by the upsurge of new hopes young people demonstrated there.[3]

In 1926 Tagore was invited to meet the Duce whose personality impressed the poet. When Rabindranath conversationally spoke of his revulsion for violence, the dictator echoed the poet's statement, saying he too abhorred violence. Krishna Kripalani, Tagore's biographer, recorded how very impressed the poet was with Mussolini. This distressed Romain Rolland who was amazed at the mendacity of the Duce's statement to the poet that he too hated violence. In the late 1920s Indian nationalist opinion was largely curious about the fascist experiment in Italy. Some Indian leaders saw it then as a country, which under a parliamentary democracy had acquired the reputation of being notoriously lax and ill governed, was being comprehensively reformed under a charismatic leader.

Nationalist Indians who wrote on Italy during this phase displayed sympathy for Italy's nationalist aspirations to make Rome again the centre of cultural exchanges between East and West. Mussolini's fascism, unlike Hitler's Nazi doctrine, was initially less humiliating to the coloured peoples of Asia because Italians spoke less on the subject of race. The German Nazis later elevated racism to a state policy to the discomfort of those Asian nationalist who for a while hoped to find in the growing Nazi clout in Europe a foil to the western imperial powers.

Romain Rolland venerated Gandhi as a contemporary St. Francis of Assissi, one who with the living faith of the early Christians was resisting pervading evils by accepting self-suffering and at the same time teaching the oppressed to resist in the like manner by non-violent non-acceptance of tyranny and by accepting poverty and self-suffering. Despairing of finding a non-violent answer to "the wind of violence" sweeping Europe, Romain Rolland found in the old crumbling world of western civilization "no refuge, no hope, no great light". The Church did not want to antagonize the mighty. Violence is whetted by national pride and "the idolatrous ideology of revolution". Pacifists prayed languishingly, as though they had no living faith in what they said. In this sombre situation Rolland looked up to Gandhi to rekindle faith in spirituality again, being able to stand up to evils.

Although in the 1930s Rolland defended the former USSR as a state keeping alive hope for the future, in 1924-25 he was distressed at the way "Moscow Communists (or those who follow their line)" were subjecting the name of Gandhi to disinformation and abuse. Communists, on the one hand, treated Gandhi as an enemy and proclaimed the bankruptcy of non-violence in India, he wrote. At the same time they "disguise Gandhi as a Bolshevik" and spread the rumour of his visit to Moscow. Gandhi's statements that if Bolshevism believed in violence and denied God, it would repel him, infuriated "the Indian Bolsheviks in Moscow."[4]

As the European crisis deepened, the French savant more than once invited Gandhi to visit Europe and show how tyranny was to be resisted and peace defended by non-violent means. In 1928, Gandhi was close to a sojourn to Europe which he cancelled as the Congress prepared for another Civil Disobedience Movement. In December 1931, Gandhi was on his way home after the failure of the second Round Table Conference to which he had rushed in terms of the Gandhi-Irwin Pact of 5 March 1931. Gandhi spent nearly six days, 5-11 December 1931 at Villeneuve Villa Olga, Switzerland, as Romain Rolland's guest. Rolland was by then a strong protagonist of the former Soviet Union; and his anxiety focused on the effete plutocratic democracies of Britain and France, and the growing menace of fascist Italy and the crisis in Germany which was soon to pass under control of the Nazis. Romain Rolland opposed Gandhi's planned visit to Italy and his meeting arranged with Mussolini in Rome. Fearing a repetition of what happened when Tagore met Mussolini in 1926 and how the fascist press exploited that event, Rolland wished Gandhi had declined the Italian invitation. But he did not want to explicitly tell Gandhi not to go. In any event, all arrangements had been made for Gandhi's travel across Italy; after meeting Mussolini, to Brindisi where he was to board the ship for India.

Still Rolland warned him of the dangers awaiting him in fascist Italy, of "underhand attempts to annexe him to their cause, as they did with Tagore..." Gandhi decided to stick to the programme so as to "see the people and bear them the mission of peace." In Rome, he would meet Mussolini if the latter wanted; but he wanted very much to meet the Pope, bearing in mind the Roman Catholic bishops in India. Rolland was not convinced. Before Gandhi left Villeneuve, Rolland had recognized that their paths were different although they probably sought the same object.

Arriving in Rome on 12 December 1931, Gandhi discovered that the Pope would not meet him but Mussolini would the same evening. Romain Rolland got reports from Madeleine Slade (Mira Behn), Mahadev Desai, and General Moris, who was Gandhi's host in Rome. They were present when Mussolini met Gandhi. However, the letter Romain Rolland got from Gandhi written aboard the ship taking him to India and to another spell of imprisonment caused Rolland distress. Gandhi's letter of 20 December 1931 said inter *alia*:

> "I had your cable. I shall get the letter referred to by you.
> "You will please write to Tolstoy's daughter (Sukhotin Tolstoy who met Gandhi in Rome on 13 December 1931) and satisfy her as to Bolshevism...

"Mussolini is an enigma to me. Many of the reforms he has made attract me. He seems to have done a great deal for the peasantry. Of course the iron glove is there. But allowing that force is the basis of Western society, Mussolini's reforms deserve an impartial study. His care for the poor people, his opposition to over urbanisation, his attempt to bring about coordination between capital and labour seem to me to demand very careful attention ... My own fundamental doubt of course abides in that these reforms are forced. But that is true even of democratic institutions. What strikes me is that behind Mussolini's ruthlessness is the motive of serving his people. Even behind his bombastic speeches there is a ring of sincerity and burning love for his people. It also seems to me that the bulk of the Italians like Mussolini's iron rule ... Needless to say I do not propose to write publicly just now about these matters. I have simply put these things before you as one knowing infinitely more than I do of them?"[5]

Unlike Tagore, Gandhi was not impressed by Mussolini's personality and remarked on the dictator's moving, wary eyes and concluded: "In general, he does not look like a man of humanity. But I must say that he was charming with me." Romain Rolland was shaken by Gandhi's letter of 20 December 1931 and regretted that he had not stopped Gandhi's travels in Italy. On 2 January 1932 he refuted the opinions Gandhi had formed of Mussolini's reforms, but did not mail it because the former was re-arrested soon after his arrival in India. However, Gandhi's letter to Romain Rolland showed that even in December 1931 it was possible for an Indian leader as percipient and observant as Gandhi to form on Mussolini's reforms favourable opinion. And Gandhi had been forewarned by no less a person than Romain Rolland whom the former respected and loved. In spite of this, Gandhi's first reactions were very different from those Rolland had explained to him in long discussions at Villeneuve.

A leader of the Indian freedom struggle, accustomed as he was to the unconcealed brutalities and economic spoliation inseparable from the British rule over India, was likely to be slow in appreciating the fine distinction Romain Rolland and several other anti-fascist intellectuals from Europe's old democracies were making in the quality of violence employed by the western democracies and fascist Italy. Rolland was peeved at Gandhi's comment that "force is the basis of Western society" and felt that he too was making the error which he had noticed in the statements of the other Indian leaders visiting him. Rolland was surprised and pained that the Indians found little difference between the violence they suffered daily in British-ruled India and the force employed by Mussolini's fascist state. Romain Rolland had reason to question the validity of Gandhi's wider conclusion that force is the basis of Western society. Rolland would have been justified in stating that force is the final recourse of all states, occidental or oriental.

Gandhi did not make public the favourable impression he had formed of certain reforms of Mussolini. It has been said that the first public reference to fascism and municipal socialism made by Subhas in September 1930 in his inaugural speech as the Mayor of Calcutta caused the Comintern to form an adverse estimate of him.

But this statement would not explain why an important Comintern official Virendranath Chattopadhyaya who together with German communist leader Munzenburg were to conduct the affairs of the League against Imperialism till Hitler took over as the Chancellor of the German Reich, was warning Jawaharlal in 1928 that he should beware of Bose.

Was it then a mere intrigue to fan rivalry between the two most promising upcoming young leaders of the nationalist camp? Or the Comintern had, on the strength of its other sources of information, concluded that Subhas's ardent nationalism and commitment to socialism whose inspiration he traced to Vivekananda and C.R. Das, would make him amenable to any proposition that subordinated his thought action to directives issued by a centre abroad claiming to proletarian international solidarity and the might of the first Soviet state? What Subhas said on fascism and municipal socialism in 1930 could not have influenced Chattopadhyaya's estimate of Subhas in 1928 when Bose had just returned to active participation in the freedom struggle after a long spell of imprisonment in Burma (Myanmar) and illnesses.

In July 1931, the Indian communist group, obeying the left-sectarian line of the Comintern's sixth Congress, stormed out of the AITUC. Subhas Bose, the elected president of the year, in July 1931, deprecated the split caused by "Moscow Communists". In the 1920s, before Romain Rolland in his despair about West Europe revised his early distaste for Bolshevik practices, he too was in the habit of describing the Comintern followers as "Moscow Communists". But in 1931-32, with fascism triumphant in Italy and, seeing the imminence of the fall of the Weimar republic in Germany and the rise of Hitler, Rolland was aggrieved that not even Gandhi could recognize the wickedness the fascist state represented.

In the letter of 2 January 1932 which he did not mail, Romain Rolland was "wonder struck" that Gandhi should have written (in his letter to Rolland) of "the good Duce, protector of the people against the exploitation of the rich!"[6] Gandhi did not use for the Duce the words Romain Rolland attributed to him. But in his annoyance Rolland omitted to repeat the more cautious words in which Gandhi had described Mussolini. It was not as though Gandhi had been unaware that violence and racism were often the preferred expressions of Nazi-Fascist gangs of the period. Dr Rajendra Prasad, one of Gandhi's close colleagues, who met Romain Rolland in August 1928 at Rigi-Kaltbad, had been so severely beaten up by a gang of Austrian Nazis at Graz that the assault left a deep wound on his forehead, and his hostess, a teacher at Graz, was attacked so violently that she fell unconscious. Dr Prasad went to Vienna for the World Congress of the International League of War Resisters, "and on the way back, at Graz in Austria, Prasad was half killed ... by Nationalists Austrian Fascists, who saw him as the representative of Gandhi."[7]

However, Gandhi, to the dismay of European anarchists and pacifists, equated the fascist state with all other states, democratic, communist or authoritarian, states which raised armies and were ready to employ military means in pursuit of their policies. His statements on these lines made at Lausanne in December 1931 shook his pacifist and anarchist European admirers. The speech so shocked the government of Switzerland that it thought of asking him to leave the country at once.[8] On the question of force as the vehicle of change or resistance, Gandhi and Romain Rolland agreed to disagree. That was why Romain Rolland finally discovered, after five days' talks, that neither on the issue of the USSR nor fascist Italy could he wholly agree with Gandhi. The latter made no distinction between Bolshevik and fascist states if both used force; for "even the worst states, 'those of Nero or Mussolini', always have something good in them and that is no reason for not rejecting them as a whole."[9]

What Gandhi wrote on Mussolini should have been read in the light of what he had said at Lausanne on the nature of all states. By 1931, before Gandhi came to Villeneuve as his guest, Romain Rolland had felt that for true Christians of Europe, whose fundamental creed demanded that they resist the evil fascist tyranny and also demand social justice for working people, Gandhi or Lenin offered them alternative forms of action. By 1930-31, Rolland's support for Gandhi was not unqualified. He waited for Gandhi to side more positively with working people against plutocrats. He was ready to allow Gandhi time, because he was convinced that the all-class front against British rule Gandhi had organised would presently make way for full-fledged Gandhian crusade for social justice, once Britain's grip on India loosened. In December 1931 at Villeneuve, Rolland explained what the dictatorship of proletariat meant "the minority of perceptive energetic workers who understand the situation feel justified in forcing masses to this unity ... in the interests of the proletariat mass whose sup it obtains by force". Gandhi rejected it. He spoke of his misgivings at the Bolshevik experiments in Russia because the basis was force.

"Just now it seems to be working well ... But I do not know how long that force is going to be effective in keeping that society, that country to this narrow path. The Indians who are under the influence of Russian methods are betraying intolerance of an extreme type. The result is that those who are under it are under a system of terrorism. So I follow the Russian experiment with a fundamental distrust", Gandhi told Rolland.[10]

Even before this, on 9 December 1931, Rolland wrote in his diary that, though his veneration and love for Gandhi remained undiminished, "I must admit that on that particular day I had the feeling that Gandhi's way is so clearly marked out before him and —in many ways—so different from mine that we have very little to discuss together; we each know exactly where we are going, and Gandhi's way is perfect for himself and his followers... But what have we to say to each other?"

Rolland, delighting in the uncompromising lucidity in Gandhi's attitude to states, is unable to act on this perception. India's non-violence was impractical in the Europe in a crisis and he had no conscientious objection to countervailing force the Soviet state employed against reaction. Rolland had no sympathy for the Indian Marxists who came to Villeneuve to portray Gandhi as a social reactionary and a friend of India's big business.

Jawaharlal Nehru, introduced by Gandhi's letter, first met Romain Rolland on 20 May 1926 at Villeneuve when he was in Europe for the treatment of ailing Kamala Nehru. Rolland's entry of that date on his conversation with Nehru on Gandhi sharply contrasts with what Jawaharlal said a year later on Gandhi. Under the May 1926 entry, Jawaharlal dwelt on Gandhi's unquestioned moral authority among all classes of Indians and the contrasting personalities of his father, Motilal, who was then the leader of the Swaraj Party, and Gandhi.

Romain Rolland noted in his diary on 1 May 1927, after another visit of Jawaharlal, Kamala, Vijaya Lakshmi Pandit and Indira:

"He seems to be breaking from Gandhism and also claims that the popular classes, workers and peasants are also breaking away ... because they see that Gandhi is doing next to nothing (Nehru says) to improve their material condition; he will hear nothing of class conflict."

After this talk Rolland concluded:

"As far as I can judge, Nehru over the last two years has moved some distance away from the religious and moral side of Gandhi's doctrine."[11]

Jawaharlal's talks with Rolland in 1926-27 mirrored the changes he underwent during his twenty-one months in Europe when he joined the League against Imperialism and visited the Soviet Union. Later, when Saumyendranath Tagore, then an adherent of the Comintern, in talks with Rolland articulated the prevailing communist line that Gandhi represented the reactionary Indian bourgeoisie and then proceeded to hyphenate Gandhi with Mussolini, the savant protested: "No, No! You cannot mention these two names in the same breath."[12] Again, an extract from Rolland's diary entry 15 October 1937 showed him disturbed by a talk he had had with Mahmud uz Zafar Khan, "and his pretty wife, Dr. Rashid Jehan". They had come armed with a letter from Jawaharlal, and, Rolland described Mahmud uz Zafar as Nehru's secretary. (Mahmud uz Zafar Khan was the Communist spokesman in the Deoli Camp in 1940-1941 when the Nazi-Soviet Pact was in force and Hitler had not invaded Russia). Rolland wrote:

"It is most curious, and rather annoying, that these elite Indian who are, in a sense, Nehru's right arm, give evidence of a sharp distaste for Gandhi ... I cannot believe that Nehru, so respectful and affectionate towards Gandhi, would approve of what they say, but he cannot be unaware of their opinion ... In addition there is in the Mahmuds a religious, may be a racial animosity. They present Gandhi as being hostile at heart to the Mohamedans ... secretly opposing. They claim that all Muslim India is against him ... They have to admit that the immense Hindu masses are with him, Mahmud says with a smile: 'Mr Gandhi is a typical Hindu.' "[13]

In 1931, when Gandhi travelled to Europe and met Mussolini, Subhas had not had any personal knowledge of fascist Italy. In 1930-31 he was in and out of prisons, travelling and addressing large public meetings in Bengal districts and other parts of the country when he could travel between the release and re-arrest. His information on the European developments of this period was drawn from newspapers, journals and books. Jawaharlal had spent twenty-one months in 1926-27 in Europe. Subhas had not been to Europe since he resigned from the ICS and left Cambridge in 1921 to join the Non-Cooperation Movement. When Subhas was at Cambridge, the place had not become the nursery of undergraduate communism which in the 1930s it was. In 1920-21, immediately after the First World War, Cambridge uncharacteristically was inclined to be conservative and student opinion was less receptive to socialism, pacifism and the stand of war resisters. Subhas was admitted to Fitzwlliam Hall. He did not join Trinity, known then as the rich brats college, which in the 1930s, as a result of the exertions by a Trinity economics don, Maurice Dobb, became the centre of communist student activities in Cambridge."[14]

Bose had been called a believer of Fascism. This assumption arose for the reason that he wanted to associate India's freedom struggle with the Second World War politics and sought the help of Fascist Italy and Nazi Germany. He also showed his admiration for Mussolini. Besides in a chapter: 'A Glimpse of the Future' in his book, *The Indian Struggle*, he, wrote: "Inspite of the anti-thesis between Communism and Fascism, there are certain traits common to both. Both, Communism and Fascism believe in the supremacy of the state over the individual. Both denounce parliamentarian democracy. Both believe in one party rule. Both believe in the dictatorship of the party and in the ruthless suppression of all dissenting minorities. Both believe in a planned industrial reorganisation of the country. These common traits form the basis of new synthesis is called *Samyavada*. Bose asserted that it would be India's task to work out this synthesis. Besides his advocacy of this synthesis, Bose had said in 1941 in Kabul that dissensions in the country would disappear only when an iron dictator rules in India for twenty years. He asserted that at least after the end of British rule in India there must be dictatorship. No other constitution can flourish in the country. And it is India's good that she should be ruled by a dictator to begin with. None but a dictator can wipe out such dissensions. He believed India needs Kamal Pasha to cure many of its ailments.

Bose had advocated dictatorship as the panacea for India's problems. However, he could not be branded as fascist. He never sanctioned the extreme tenets of fascism which advocated imperialistic expansion and believed in racialism. He was a votary of the rights of the exploited masses. Thus, although he took armed help from the fascist powers of Europe and Asia, and organised the Indian National Army for India's liberation, yet he did not practice the ideology of fascism. For the Forward Block, the party he founded, he summarised the dominant guiding principles on January 1, 1941:

(1) complete national independence and uncompromising anti-imperialist struggle for attaining it;
(2) a thoroughly modern socialist state;
(3) scientific large scale production for the economic regeneration of the country;
(4) social ownership and control of both production and distribution;
(5) freedom for the individual in the matter of religious worship;
(6) equal rights for every individual.
(7) linguistic and cultural autonomy for all sections of the Indian community; and
(8) application of the principle of equality and social justice in building up New Order in Free India.

The fascist overtones within his perception of socialism, also revealed Subhas's peculiar attitude towards communism. He regularly talked of a synthesis of Russian communism and Italian and German fascism. The impression that one gets is that he was more influenced by the powerful authority of Stalin, Mussolini and Hitler, than their specific ideologies and class positions. And although in 1938, in his interview with Rajani Palme Dutt, he rejected his own earlier suggestion about "a synthesis between Communism and Fascism," as it was based upon a wrong perception of Fascism.[15] Even in 1940, Subhas advocated, as his party's mouth piece, "Forward", a political path which would follow simultaneously the ideal of Lenin's 'wisdom', his 'dare devil attempt' and Mussolini's 'daring and uncompromising policy', with the path of the 'Irish Sinn Fein revolution.'[16] These are clear signs of politics which sought to combine, 'social good' with 'national pride' in an independent nation state—an independence which was to be achieved through violent means. It was simultaneously, a recognition of the necessity and desirability of a modern 'economically powerful' and centralised nation state, and that of the impossibility of achieving it through 'peaceful' (i.e. hegemonic) political organisation.

That Subhas was one of the most ardent supporters of a 'state capitalist' economic model is well documented. Indeed, Subhas was instrumental in setting up of the National Planning Committee in October 1938,[17] whose main objectives were to plan the development of heavy industries and the infrastructural sector of the economy, in order to protect and encourage the national bourgeoisie. It is precisely on the question of the development of private capital that Subhas believed that Gandhi would fall in line. Three years earlier, he had written (in 1935):

"No doubt, the Mahatma has condemned the 'machine civilisation' of the modern world and has eulogised the good old days when men were content with their cottage industries ... But that is a personal belief or idiosyncrasy. Whenever he has expounded the contents of Swaraj, he has spoken in the language of mid-Victorian parliamentarian democracy and of traditional capitalist economics."[18]

Subhas's own politics also broadly fell in line with the interests of the national bourgeoisie, although at times he veered towards more radical positions, especially during the days of his connections with the Left. Yet, unlike Jawaharlal Nehru, he was not a liberal bourgeoise democrat of the Western type. In fact he had learnt very early in his political career that liberal democratic forms of political practice were unsuited for a country like India, where the institution of citizenship had not yet developed. Further, the liberal democratic notions of the individual-citizen's interests being represented in the state required a more homogenous society than India. Thus, for Subhas, dictatorship was the most suitable state form, whether it be personal or that of the party which would win Swaraj for the nation.[19] It is only through a strong central government that the differences between the people could be ironed out[20] and more importantly, "order and public security can be safeguarded."[21] Subhas continued,

"Behind this government will stand a well organised, disciplined all-India party, which will be the chief instrument for maintaining national unity."[22]

Subhas also wrote: "I tried to study something of the work of the Fascist Party...... I could see and feel that the Fascist party was out to create a new nation according to its ideas and ideals."[23] But to him more relevant was the Party organisation, not the ideals. He continued, "What was of greater interest was not the ideal of fascism but the thoroughness of the party organisation. We have our national ideals which we do not need to borrow from elsewhere. But what we do lack is organisation and discipline."[24]

With this we come directly to the question of the character of the political party that Subhas had envisaged in order to lead the freedom struggle. This party was going to be an alternative to the Indian National Congress and was to be organised along the principles of military discipline. It was to be opposed to 'mid-Victorian' democratic politics and would try to fight for the establishment of a strong Centralised government with dictatorial powers.[25] Alongwith militaristic discipline, the party was also to lead an armed struggle of the people in order to establish a new state. The armed struggle was also to be aided by a revolt in the army.[26]

The motifs of single party dictatorship, military discipline and armed struggle, were definitely the most important elements that Subhas borrowed from the communist movement and national socialism. While Gandhi too had stressed the necessity of discipline within the party and talked of the *satyagrahi* as a disciplined soldier,[27] this discipline was one of non-violence. Subhas' proposed party, however, was to stand outside these forms of passive resistance. The party was to organise an underground armed struggle by violently disrupting the functioning of the colonial state.[28]

This plan for a new party, marked a definite break with all forms of hegemonic politics, especially that of the Congress. The new party was not to be dependent on mass support nor was it to be based on mass mobilisation. Instead, it was to be a party of a band of dedicated and disciplined 'missionaries'[29], devoted to the cause of freedom. It was this militaristic party which was to lead the masses and was to take over power in order to discipline the masses themselves. For the masses were 'unsophisticated' and ridden by differences of caste, religion, region, etc., all of which were barriers to the development of a strong nation. The politics of persuasion had only succeeded in throwing up leaders such as Gandhi, who were ill equipped to transform India into a modern nation. It was only a radical break with this politics of passive persuasion that could modernise India. Archives which have recently been made public seem to confirm our thesis that Subhas was opposed to the politics of ideological persuasion or hegemony. It now appears that "Bose believed that the country should be subjected to an authoritarian regime for a period of 20 years in which time he assumed that the tortured social and economic equations which in effect, continue to interfere with harmonious parliamentary practices would have sorted themselves out."[30]

However, Subhas' dream of an armed struggle was to take a totally different shape from what he had imagined in the late 30s. With the intensification of the War in Europe, Subhas decided to strike British Imperialism at the moment of its weakness. The rest, of course, is history and an often repeated one at that. Although the I.N.A. never became a force strong enough to seriously challenge the British state, it had a profound impact on the morale of the Indian people. The notion that one of their own, was fighting the powerful British Empire, was a source of immense strength to many a freedom fighter. The I.N.A. trials drew a lot of attention and aroused tremendous public sympathy. Indeed Jawaharlal Nehru, who was otherwise not known for any sympathy for Subhas Bose, became one of the defence lawyers for the accused. The R.I.N. mutiny too, drew inspiration from the valiant struggles of the I.N.A., and the fear of its spirit spreading into the Army, became a cause for serious concern for the Colonial State.[31]

REFERENCES

1. Burham, James, *The Managerial Revolution* (England: Penguin Books Ltd.), 1942.
2. *Romain Rolland and Gandhi Correspondance* (New Delhi: Publications Division, Government of India), 1976, p. 67. Extract from Rolland's diary entry of 5 September 1926.
3. Pandit, V.L., *The Scope of Happiness* (New Delhi: Vikas Publishing House), 1965.
4. Rolland, Romain, Mahatma Gandhi, translation from French by Catherine D. Groth (New Delhi: Publications Division, Government of India), pp. 41-42.
5. *Romain Rolland and Gandhi Correspondance, op. cit.*, pp. 241-242.
6. *Ibid.*, p. 247.
7. *Ibid.*, Rolland's letter to Toshihiko Katayama, Japan, of 15 August 1928, pp. 415-416.
8. *Ibid.*, Rolland's letter to Pastor Paul Leenchardt of 20 December 1931, pp. 449-450.
9. *Ibid.*, Rolland's Sumary of Gandh's Lausanne Speech, p. 432.
10. *Ibid.*, pp. 225-230.
11. *Ibid.*, pp. 59-60.
12. *Ibid.*, p. 291.
13. *Ibid.*, p. 343.
14. Seale, Patrick and Maureen McConville, P. (London: Penguin Books Ltd.), 1978, pp. 50-51.
15. Bose, Subhas Chandra, *The Indian Struggle, op. cit.*, p. 393.
16. Sarkar, M.L. (ed.), *Subhas Bose: 1939-40* (Bengali), Translated by Nipen Chakravarty (Calcutta: Forward Publishing Concern), 1975, pp. 78, 90 and 120.
17. *Ibid.*, p. 360.
18. Bose, Subhas Chandra, *The Indian Struggle, op. cit.*, p. 316
19. *Ibid.*, p. 386.
20. *Ibid.*, p. 312.
21. *Ibid.*, p. 454.
22. *Ibid.*
23. Bose Subhas Chandra, *Impression in Life* (A Collection of his articles and press statements published in Europe) (Lahore: Hero Publication), 1947, p. 20.

24. *Ibid.*, p. 21
25. *Ibid.*, p. 312.
26. Sarkar, M.L. (ed.), *Subhas Bose: 1939-40* (Bengali), Translate by Nipen Chakravarty, *op. cit.*, pp. 113-114.
27. Guha, R. (ed.), *Subaltern Studies III, op. cit.*, p. 156.
28. Even in 1933, 'armed struggle' had been a viable option in Bose's mind, but he had rejected it on the grounds of its incompatibility with the Congress pledge of non-violence. See Presidential Speech at the third Indian Political Conference held in London. *The Indian Struggle, op. cit.*, p. 367.
29. *Ibid.*, p. 377.
30. "The Proto-Dictator" Editorial, *The Steaman*, Delhi, 3 September 1997.
31. Sarkar, S., *Modern India: 1885-1947* (Delhi: Macmillan), 1983, p. 124.

7

A CRITIQUE OF GANDHIAN LEADERSHIP

Gandhi was the most influential personality of Indian freedom movement. All the major initiatives from 1920 to 1945 emanate from him. Subhas Bose was also influenced and inspired by him appreciated Gandhiji in two ways:

(1) Offered his comments on the leadership of Gandhiji in a general way. He has highlighted the positive as well as the negative points of Gandhian leadership.

(2) Offered his critical observations on some definite movements and actions of Gandhiji.

Positive Aspects of Gandhian Leadership

Subhas Bose had devoted a full chapter (Chapter 16) in the book *The Indian Struggle* (1935) to evaluate the role of Mahatma Gandhi in Indian history.

Subhas thought that the role of a man in history depends upon two factors. His physical and mental equipment and the environment and the needs of the times in which he is born. There was something in Mahatma Gandhi which appealed to the Indian masses. Born in another country like Russia, Germany or Italy, he might have been a complete misfit; his doctrines and mode of living would have led him to the cross on to the mental hospital. In India it was different. His simple life, his vegetarian diet, his goat milk, his day of silence, his habit of squatting on the floor instead of sitting on a chair, brought him nearer to his people. They thought that he was a product of Indian soil. Gandhiji spoke in a language that was understood by the Indian masses. He did not quote Herbert Spencer or Edmund Burke, instead referred to scriptures like *Mahabharata* and *Ramayana*, that Indians understood. When he used words like *Ramrajya, Ahimsa* and *Brahmacharya*, people identified with such concepts completely.

The time was also favourable to the Mahatma. Gandhiji would not have succeeded at a time of revolutionary movement like the revolution of 1857 when people took their arms and were able to fight violently. They wanted a leader who could lead them in war. Gandhiji could not provide an effective leadership in such a situation. But when Mahatma Gandhi emerged on the political scene, India was disarmed. The armed struggle against the mighty British empire could not succeed. The superior equipment of Britain could smash any armed uprising. In the words of Subhas Bose:

"In 1920 India stood at cross-roads—constitutionalism was dead; armed revolution was sheer madness. But silent acquiescence was impossible. The country was groping for a new method and looking for a new leader. Then there sprang up India's man of destiny, Mahatma Gandhi, who was biding his time all these years and quietly preparing himself for the quiet task ahead of him. He knew himself, he knew his country's needs and he knew also that during the next phase of India's struggle, the crown of leadership would be on his head. No false sense of modesty troubled him. He spoke with a firm voice and the people obeyed."[2]

The Indian National Congress during the freedom struggle was the creation of Mahatma Gandhi. The Congress Constitution was his handiwork. From a talking body he converted the Congress into a living and fighting organisation. It had its ramification in every town and village in India. Nobility of character and capacity to suffer were made the essential tests of leadership. Under the leadership of Mahatma Gandhi, the Indian National Congress was the largest and the most representative political organisation in the country.[3]

Gandhiji achieved success by his single hearted devotion, relentless will and his indefatigable labour. He was not too revolutionary for the majority of his country. His policy was one of unification. He wanted to unite Hindu and Muslim, the high caste and the low caste, the capitalist and the labourer, the landlord and the peasant. By his humanitarian outlook and his freedom from hatred, he was able to raise sympathy even in his enemy's camp.[4]

Subhas Bose has tried to identify the reasons for the success of Gandhiji. It was a known fact that Gandhiji was considerably influenced by the teachings of Jesus Christ and the ideas of Leo Tolstoy. It can not be claimed that he was altogether original in his ideas or moral in his practice. But his real merit was a two-fold one. He translated the teachings of Christ and the ideas of Leo Tolstoy and Thoreau into actual practice. He demonstrated that it was possible to fight for liberty without resorting to violence. Gandhiji used non-cooperation not for remedying local grievances, but for winning national freedom and he well nigh demonstrated that it was possible to paralyse the civil administration of a foreign government thereby and bring the government to its knees. A happy combination of factors helped to bring Mahatma Gandhi to the forefront in 1920. The attempted revolution during the Great War had failed and the Revolutionary party had been crushed. There was consequently no possibility of another revolution in 1920. The country wanted a bold and vigorous policy on the part of the Congress and the only alternative was a movement like the one which Mahatma Gandhi started. The death of Lokamanya Tilak on the eve of the Calcutta Congress removed from the field the only leader who could challenge Gandhiji. Thus Gandhiji was fully prepared to assume the undisputed leadership of the Indian National Congress in 1920. Through strict discipline he had equipped himself for a life of suffering and during the apprenticeship in Indian politics, from 1914 to 1920, he had been able to gather round him a band of loyal and trusted followers. Among the prominent leaders who joined Mahatma Gandhi were C.R. Das from Calcutta, Pandit Motilal Nehru and his son Jawaharlal Nehru from Allahabad, Pandit Govind Ballabh Pant from Nainital, Lala Lajpat Rai from Punjab, Vitthalbhai Patel and Vallabhbhai Patel from Bombay, N.C. Kelkar from Poona, Dr Moonje and Mr Abhyankar from Central Provinces, Rajgopalchari, A. Rangaswami Iyenger and Satyamurthi from Madras, Mr Prakasam from

Hyderabad, Dr. Rajendra Prasad from Bihar, Dr. Husain from Delhi. Maulana Abul Kalam Azad, Maulana Shaukat Ali and Maulana Mohammed Ali were the eminent Muslim leaders who supported Mahatma Gandhi. Many well to do lawyers gave up their professional work in response to the appeal of the Congress.

Gandhiji made use of Satyagraha in South Africa as well as in India a number of times with very good results. Lastly, Gandhiji had gathered round his head a halo of saintliness. This was of considerable value to him. Gandhiji emerged as the virtual dictator of the Congress.

The Non-cooperation Movement gave the country a highly organised party organisation. Before that the Congress was a constitutional party and mainly a talking body. Mahatma Gandhi gave it a new constitution and a nationwide base. He converted it into a revolutionary organisation. The tricolour national flag red, green and white was adopted all over the country and assumed great importance. Uniform slogans were repeated everywhere and a uniform policy and ideology gained currency from one end of India to the other. The English language lost its importance and the Congress adopted Hindi as the common language for the whole country. Khadi became the official uniform for all Congressmen. In short, all the features of a modern political party became visible in India. The credit for such achievements belonged to the leader of the movement, Mahatma Gandhi.[5]

The trial of Mahatma Gandhi was an historic event. In describing the trial proceedings, C.R. Das in his presidential speech at the Gaya Congress in December 1922, drew an analogy with the trial of Christ before Pontius Pilate.[6] In the trial Mahatma Gandhi made a lengthy statement, describing how from a staunch loyalist and co-operator he had become an uncompromising disaffectionist and non-co-operator. He ended his statement with these words:

"The only course open to you, the Judge and the Sessions, is either to resign your posts and thus dissociate yourselves from evil if you feel that the law you are called upon to administer is an evil and that in reality I am innocent, or to inflict on me the severest penalty if you believe that the system and the law you are assisting to administer are good for the country and that my activity is therefore injurious to the public work."[7]

The English Judge Mr. Broomfield, sentenced Gandhiji to six years imprisonment.

On February 27th, 1930 Mahatma Gandhi announced his plan of campaign. The next four moves taken by him stand out for all time to come as some of the most brilliant achievements of his leadership. They reveal the height to which his statesmanship could ascend in times of crisis. Mahatma Gandhi made it clear that the movement would be non-violent but it would not be stopped so long there was a single civil resister left free or alive.[8]

Gandhiji's march to Dandi was an event of historical importance which would rank on the same level with Napoleon's march on his return from Elba or Mussolini's march to Rome when he wanted to seize political power.[9] Fortunately for the Mahatma, he had a wonderfully good press within India and outside. In India, for days and days, every detail connected with the march found the widest publicity. The march on foot enabled him to rouse the entire countryside through which he passed. It also gave him time to work up the feelings of the country as a whole. If Gandhiji had travelled by train, he could not make such an impact. While Gandhiji was marching from village to village, an intense propaganda was carried on asking the people to give up service under the British Government. At every step Mahatma Gandhi received warm welcome.

He appealed to women of India to join the civil disobedience movement. The appeal was transmitted all over the country. It had a magical effect; even the women of the most orthodox and aristocratic families went to prison.[10]

Whatever may be the permanent value of the Poona Agreement, there is no doubt that the fast of Mahatma Gandhi had a permanent and far reaching effect in rousing the conscience of the Hindu community. It was a unique spectacle to see how the heart of the entire nation throbbed for one man. All sections of the Hindus were stirred to activity as they had never been before. The most important result of the epic fast was to give a powerful impact to the movement for the eradication of untouchability.[11]

Subhas Bose made moving references to Mahatma Gandhi. While concluding his Presidential address at the Haripura Congress in February 1938, he hoped and prayed that Mahatma Gandhi might be spared for our nation for many years to come. India needed him to keep its people united and to keep the freedom struggle free from bitterness and hatred. India was fighting against world imperialism. India freed meant humanity saved.[12]

The Gandhian Movement: A Synthesis

Subhas Bose wrote an essay on Forward Block[13] in January 1941 during his secret sojourn in Kabul. He tried to interpret Indian nationalist movement in terms of dialectical process involving thesis, anti-thesis and synthesis. Subhas Bose considered the colonial rule as the thesis, liberal phase of Indian politics as anti-thesis and the Gandhian movement as the synthesis. In 1920 Gandhi took possession of the Indian National Congress and for two decades he maintained his hold. This was possible not merely because of Mahatma Gandhi's personality but also because of his capacity to assimilate other ideas and policies. During its twenty years of domination of the Congress, whenever revolts appeared, the Gandhi movement took the wind of their sails by accepting many of their ideas and policies. When the Swaraj Party arose in 1923, the conflict that followed continued only for a time. At the Cawnpore Congress in 1925, the Swarajist policy of carrying non-cooperation inside the legislations was accepted by the Gandhiites and was thereupon adopted by the Congress as a whole. Again in 1928, at the Calcutta Congress there was a revolt against Gandhism on the issue of Independence. Mahatma Gandhi then advocated Dominion Status. Subhas Bose moved an amendment to the resolution of Mahatma Gandhi and called for complete independence. His amendment was defeated. But a year later at the Lahore Congress Mahatma Gandhi himself moved the resolution for complete independence.

By this process of assimilation, the Gandhi movement was able to maintain its progressive character and prevent the emergence of any big left wing movement. There was a temporary set back after the Gandhi-Irwin Pact in March 1931, but Gandhiji recovered lost ground when he launched civil disobedience movement in January 1932.

The Azad Hind Fauz founded by Subhas Bose crossed the Indo-Burma frontier on 18 March, 1944. It carried the armed struggle on to Indian soil. Subhas Bose in his address to Mahatma Gandhi on November 6, 1944 showed his deep anxiety for Gandhiji's health. Subhas gave a detailed justification for his course of action. He paid handsome compliments to Gandhiji's leadership and sought the blessings and the good wishes of the Father of Our Nation in the holy war that was then raging in north eastern India. He confessed that all the countries free from British influence that he had visited, Gandhiji was held in the highest esteem as no other Indian political leader had been during the last century. The high esteem in which Gandhiji was held by patriotic Indians outside India and by foreign friends of India's freedom, increased hundred fold when he bravely sponsored the Quit India resolution in August 1942.[14]

Negative Aspects of Gandhian Leadership

While appreciating some positive aspects of Gandhian leadership, Subhas was nevertheless critical of some negative aspects of Gandhiji's stewardship.

Subhas had his first meeting with Mahatma Gandhi on 16 July, 1921 at Mani Bhavan in Bombay. Subhas asked a number of questions regarding the Non-cooperation Movement of Gandhiji. He did not feel satisfied with the replies of Gandhiji. He felt that there was a deplorable lack of clarity in the plan which Gandhiji had formulated. Mahatmaji himself did not have a clear idea of the successive stages of the campaign which would bring India to his cherished goal of freedom. Gandhiji had promised freedom in a year. It was not at all clear how could Britain agree to grant India independence within such a short time. To Gandhiji it was a question of faith. Possibly Gandhiji expected a change of heart on the part of the British Government leading to an acceptance of India's national demands.[15]

The Treaty of Severs published in the summer of 1920 gave rude shock to Indian Muslims. The Turkish empire was dismantled and the Sultan of Turkey was humiliated. Mahatma Gandhi espoused the Muslim cause and started the Non-cooperation Movement. One of the objects of the movement was redressal of Khilafat. The introduction of the Khilafat issue in Indian politics was unfortunate. Under Kamal Pasha, Turkey emerged as a secular republic. In 1922 the institution of Khilafat was abolished and the Khalifa was exiled.[16]

Subhas Bose did not find any fault with Mahatma Gandhi for his intimate association with the Ali Brothers or connecting the Khilafat issue with the other national issues. But Gandhiji made a mistake by allowing the Khilafat Committee to be set up as an independent organisation throughout the country, quite apart from the Indian National Congress. The result of this was that when Kamal Pasha later forced the Sultan to abdicate and abolish the office of the Khalifa, the Khilafat question lost all significance. The majority of the workers of the Khilafat organisation were absorbed by sectarian, reactionary and pro-British Muslim organisations. If separate Khilafat Committees had not been organised and all the Khilafatist Muslims had been persuaded to join the ranks of the Indian National Congress, they would have been absorbed by the Congress when the Khilafat issue became a dead one.

Mahatma Gandhi suspended the Non-cooperation Movement in February, 1922 as a result of the Chauri Chaura tragedy in a U.P. village near Gorakhpur. The police had opened fire on a procession but when their ammunition was exhausted, they shut themselves up inside the building. The excited mob set fire to it. The members of the police force, twenty-two in number, were burnt to death.

The sudden suspension of the movement caused much resentment among important Congress leaders, who took Gandhiji to task. Bose wrote, "To sound the order of retreat just when public enthusiasm was reaching the boiling point was nothing short of a national calamity".[17] Subhas Bose quoted C.R. Das: "... the Mahatma opens a campaign in a brilliant fashion, he works it up with skill, he moves from success to success till he reaches the zenith of his campaign, but after that he loses his nerve and begins to falter."[18]

The end of non-cooperation marked the beginning of reaction. Mahatma Gandhi was arrested on 4th March 1922 and sentenced to six years' imprisonment. In the opinion of the Civil Disobedience

Enquiry Committee set up by the Congress, Non-cooperation had achieved little. The movement had failed to achieve its objective of securing the redress of Khilafat and Punjab Wrongs and the attainment of Swaraj. Subhas Bose attributed the failure of non-cooperation to Mahatma Gandhiji's faulty strategy. The promise of Swaraj within one year was not only unwise but childish.[19]

The sudden stoppage of the Non-cooperation Movement brought the Congress-League alliance to an end. The Hindu-Muslim unity forged during the common struggle began to disintegrate. Towards the end of 1921, the Moplah rising occurred in Malabar with the object of establishing a Khilafat State. The fierce Moplahs slaughtered not only a few British officials but far more of their Hindu neighbours. A series of communal riots raged with brief intervals for many years. The growth of religious element in Indian politics was an unhealthy development.

Though the leaders and the rank and file of the Swaraj Party had the highest respect for the personality of the Mahatma Gandhi, the Party was an anti-Gandhi Party. It was strong enough to force Gandhiji to voluntary retirement from politics. This retirement continued till the Calcutta Congress in December 1928.

According to Subhas Bose, Mahatma Gandhi came to be looked upon by the mass of the people as a Mahatma before he became the undisputed political leader of India. Consciously or unconsciously Gandhiji fully exploited the mass psychology of the people. In some parts of the country Gandhiji began to be worshipped as an Avatar. To make matters worse, political issues were not discussed in the light of pure search. They were mixed up with ethical issues.

The Madras Congress in December 1927 passed a resolution proclaiming complete independence as its goal. Mahatma Gandhi described the resolution as hastily conceived and thoughtlessly passed. The Calcutta Congress was held in 1928 under the presidentship of Pandit Motilal Nehru. These were two groups in the Congress. The older group was content if they had a dominion form of government. The left wing addressed to the goal of complete independence. The main resolution of the Congress was moved by Mahatma Gandhi. It called for the setting up of the dominion status. Subhas Bose moved an amendment to the effect that the Congress would be content with nothing short of independence. The amendment was supported among others by Pandit Jawaharlal Nehru. The amendment was lost by 973 votes to 1350. The followers of Mahatma Gandhi made it a question of confidence. They gave out that if Gandhiji was defeated he would retire from the Congress. Many people voted for this resolution not out of conviction, but because they did not want to be party to forcing the Mahatma out of the Congress. The voting showed that the left wing was becoming stronger.

The next annual session of the Congress was held at Lahore under the presidentship of Jawaharlal Nehru. The choice of Jawaharlal Nehru as Congress president lent a special glamour to the session. Nehru was an embodiment of the youthful ardour. The resolution for complete independence was moved by Mahatma Gandhi. The resolution declared complete independence as the goal of the Congress. As a preliminary towards organising a campaign for independence, the Congress resolved upon complete boycott of the Central and provisional legislatures constituted by the government. It called upon the Congressmen and others taking part in the national movement to abstain from participating in future elections. It directed the present Congress members of the legislatures to resign their seats. The resolution appealed to the nation to carry on the constructive programme.

The Revival

Subhas Bose moved a resolution to the effect that the Congress should aim at setting up a parallel government in the country. This resolution was defeated. Subhas Bose observed that though the Congress accepted the goal of complete independence as its objective, no plan was laid down for reaching the goal. Subhas Bose was not included in the new working committee. Gandhiji said that he wanted a committee that would be completely of one mind. It became a question of the confidence in Gandhiji. As the House did not want to repudiate him, it had no option but give into his demand.

Subhas Bose has summed up the lessons of the Lahore Congress as follows:

"Altogether the Lahore Congress was a great victory for the Mahatma, Pandit Jawaharlal Nehru, one of the most prominent spokesman of the left wing, was won over by him and the others were excluded from the Working Committee. The Mahatma could henceforward proceed with his own plans without fear of opposition within his Cabinet, and whenever any opposition was raised, he could always cover the public by threatening to retire from the Congress or to fast unto death. From his personal point of view, it was the cleverest move with a subservient cabinet; it was possible for him to conclude the pact with Lord Irwin in March 1931; to have himself appointed as the sole representative to the Round Table Conference to conclude the Poona Agreement in September 1932 and do other acts which have done considerable disservice to the public cause."[20]

While the Civil Disobedience Movement was in progress, the British Government held the first Round Table Conference in London towards the end of 1930. The Congress did not participate in the Conference. Without the Congress, the Conference had been like a dramatic representation of Hamlet without the Prince of Denmark. To facilitate a compromise with the Congress, Mahatma Gandhi and the members of the Congress working committee were released in January, 1931. Mahatma Gandhi held negotiations with the Viceroy Lord Irwin and signed a pact with him on March 5, 1931. According to the terms of the Gandhi-Irwin Pact, the Viceroy agreed to (1) release all political persons including those convicted of violent offences, (2) restore confiscated property, (3) permit persons living within a certain distance of the seashore to collect or manufacture sale free of duty, and (4) permit peaceful picketting of liquor, opium and foreign cloth shops. The Congress on its part, undertook to (1) suspend Civil Disobedience, (2) give up its demand for an impartial inquiry into police excesses, and (3) participate in the second session of the Round Table Conference on the basis of Federation and Responsibility with reservations or safeguards in the interest of India.

Radical nationalists condemned the Pact as an abject surrender to imperialism. Jawaharlal Nehru felt a tremendous shock on account of the clause relating to safeguards. Indian youth was deeply stirred by the execution of Bhagat Singh and his comrades. They were indignant with Mahatma Gandhi's failure to secure pardon for them. Subhas Bose made the following criticism with regard to the Gandhi-Irwin Pact:

(1) The Pact had gone with many petty and unnecessary details but had avoided the main issue of Swaraj.

(2) The Conference was really no Round Table Conference because there was no finality about the decisions of the Conference. The whole matter was to be reconsidered by the British Parliament. In a real Round Table Conference the decisions are always final and binding on both parties. The name Round Table Conference had been used only to hoodwink foolish Indian politicians.

(3) The Indian delegates to the Conference were to be selected not by the Indian people but by the British government.

(4) The conference was not confined to the representatives of the two belligerent parties. Nondescripts of all kinds who had nothing to do with the fight for Swaraj, were to be there to throw obstacles in the path of the Indian nationalists.

(5) The proposal of a Federation between nationalist British India and the autocratic Indian Princes was an absurd one. The Princes or their nominees would act as a dead weight against the national forces.

(6) Safeguards take away what responsibility gives. It was a gross error on the part of the Mahatma to talk of safeguards in the interests of India. The only safeguard that the Indians wanted was liberty. The actual safeguards were demanded by the Britishers and were against the interest of Indians. It was wrong to induce the Indian people to accept such safeguards by saying that they were in the interests of India.

(7) The amnesty provided under the Pact was inadequate because the following classes of political prisoners were included:

(a) The state prisoners and detenus imprisoned without trial. These were about one thousand prisoners in Bengal alone, (b) Prisoners convicted of revolutionary offences, (c) Prisoners under trial for alleged revolutionary offences, (d) The under trial prisoners in the Meerut Conspiracy Case, (e) Prisoners incarcerated in connection with labour strikes and other labour disputes, (f) The Garhwali soldiers who had been court martialled and given heavy sentences for refusing to fire on unarmed citizens, and (g) Prisoners sentenced in connection with the Civil Disobedience Movement, the charges against whom referred to violence of some sort.

(8) The demand originally made by Mahatma Gandhi for an inquiry into police atrocities during the Civil Disobedience Movement was excluded from the Pact.

According to Subhas Bose, the Gandhi-Irwin Pact proved to be not a blessing but a curse. The time was not opportune for an understanding of the sort as the struggle should have been continued for some time longer. The pact contained nothing of value. These was no one in the working committee to point out the correct thing to Mahatma Gandhi. Subhas Bose felt that men with fixed ideas were not well qualified for political bargaining. So far as Gandhiji was concerned he alternated between obstinacy and leniency. He was also too susceptible to personal appeals. With such habits of mind, it was difficult to get the better of one's opponent in political bargaining. The Gandhi-Irwin Pact was a great help to the Government. It gave them time to inquire more deeply with the tactics of the Congress and those after to perfect their machinery for dealing with that body in future.[22]

In pursuance of the Gandhi-Irwin Pact, Mahatma Gandhi participated in the second session of the Round Table Conference at London as the sole representative of the Congress. Subhas Bose considered this decision of the Congress as thoroughly wrong. Alone in an assembly of about one hundred men, with all kinds of nondescripts, turnkeys and self-appointed leaders arranged against him, he had to be at a great disadvantage. He had no body at his side to back him up in the fight that he would have with the reactionary Muslim leaders. There was no help for that. The blind followers of the Mahatma were not expected to criticize him. Those who were not his unorthodox followers had no influence on him regardless of their character, wisdom, or experience.

Gandhiji reached London on September 12th, 1931. During his story in London, between September 12th and December 1st, Mahatma Gandhi spoke twelve times at the Round Table Conference.

During his stay of nearly three months in England, Mahatma Gandhi had an exceedingly busy time. He overstrained himself. Some times he did not allow himself to have more than two hours sleep for days together. He met all sorts of people there. During week ends he undertook tours to Cambridge or Oxford or Lancashire in order to awaken interest for India. There was lack of coordination and unity of purpose in all his activities. Indian members of the Round Table Conference complained that it was difficult to contact Mahatma Gandhi when they wanted him. Indian Liberal members of the Round Table Conference complained that instead of playing a lone hand, he could have rallied all the anti-communal forces and become a leader of the United Nationalist Party. The Mahatma's visit to England was badly planned, his personal entourage did not consist of any advisers worth the name. Subhas Bose has identified a deeper cause which accounted for Gandhiji's failure at the Round Table Conference. During his stay in England Gandhi had to play two roles in one persons: the role of a political leader and that of a world teacher. Sometimes he conducted himself not as a political leader who had come to negotiate with the enemy, but as a master who had come to preach a new faith, that of non-violence and world peace. Because of his second role, he had to spend much of his time with people who were quite useless in promoting his political mission.

Reviewing Gandhiji's visit to Europe, Subhas Bose regretted that he spent so much of his time in England and so little on the continent. Even on the continent he did not devote sufficient time on attention to politicians, industrial magnates and other people who really counted in politics.

The first two sessions of the Round Table Conference had failed to resolve the baffling deadlock over the communal deadlock. At the end of the second conference, the British Prime Minister Ramsay Macdonald had told the delegates that while the primary responsibility for settling the problem on the communities concerned, the British Government was determined that this disability should not have be permitted to progress. He declared that if an agreed solution was not forthcoming, the British government would be compelled to apply a provisional scheme of its own. The outcome was the Communal Award which was published on 8th August 1932. The most atrocious feature of the scheme was that the depressed classes were recognised as a distinct minority and given the right to choose their representatives through separate electorates with an additional vote in the general constituencies. The Communal Award was in keeping with the traditional British policy of accentuating communal divisions in India in order to neutralise the forces of Indian nationalism.

Mahatma Gandhi could not accept this. He went on a fast unto death to undo the communal award. The fast had the desired effect of getting the Hindu consciousness into action. Due to the efforts of the leading public men, a compromise formula was evolved which Mahatma Gandhi accepted with satisfaction and which Dr. B.R. Ambedkar signed somewhat half-heartedly. According to this formula, the depressed classes were given more seats than the Communal Award offered them. But election for these seats was to be conducted in two stages. In a preliminary election the untouchables were to select from candidates for each seat on the basis of separate electorates. But the caste Hindus and Harijans were to vote jointly in the final election. In addition, depressed classes were given an additional vote in the election for general seats not reserved for them. The agreement was known as Poona Pact. It was adopted on 26th September, 1932. The same day Gandhiji broke his fast.

Subhas has criticised the Poona Pact on the basis that it served to advertise to a disproportionate degree the issue of the depressed classes. Hitherto the world had known only one issue relating to India, the political issue—India's grievance against England. Now, Mahatma Gandhi announced to the world that there was another issue, the internal issue of such vital importance to India that he was prepared to stake his life for it. British propagandists were not slow to take advantage of the opportunity. In September 1932 the whole of Europe was told that the Mahatma was fasting because he was against granting certain rights to the untouchables.

The fast had another unfortunate effect which proved to be more serious. It served to side track the political movement at a time when all possible attention should have been devoted to it.[24]

Subhas Bose was unanimously elected the Congress president for its fifty-first session to be held at Haripura (Gujrat). It was a distinct pointer to the growing influence of the radical wing of the Congress which was not in tune with Gandhiji's ideology. The Haripura Congress met at a time when the political sky of Europe was overcast with war clouds. Subhas Bose after the Munich Pact, in September, 1938 began an open propaganda throughout India to prepare the people for a national struggle. This move was resented by the orthodox followers of Gandhiji. The fundamental difference between Subhas Bose and Mahatma Gandhi was in their attitude towards Britain. Bose looked upon the war between Britain and Germany as an opportunity for India.

The differences between Gandhiji and Subhas Bose came to a head over the re-election of Subhas Bose as the President of the Congress. Gandhiji favoured the candidature of Dr. Pattabhi Sitaramayya. Subhas won inspite of the opposition of Gandhiji. Gandhiji issued a statement on 31st January 1939 after the decisive victory of Subhas. Gandhiji said that since he was instrumental in inducing Dr. Pattabhi Sitaramayya not to withdraw, the defeat was actually his own. Thirteen members of the Congress Working Committee resigned from the body. Ultimately Subhas Bose had to resign from the Congress Presidentship. Subhas Bose organised his own party Forward Block which was not much of a success. The Congress Working Committee passed a resolution by which Subhas Bose was disqualified as President of the Bengal Provisional Congress Committee and to be a member of any elective Congress Committee for three years as from August, 1939. The outbreak of World War II on 3rd September, 1939 precipitated a grave constitutional crisis in India. England had taken up arms against Nazi Germany with the professed aim of saving freedom and democracy. The Viceroy forthwith declared India to be in a state of war against Germany without informing or consulting the representatives of her people in the Central or Provincial Legislatures. As a protest the Congress Ministers resigned in eight provinces.

Subhas wanted the Congress to start a mass movement at this junction. He even went to Wardha to persuade Mahatma Gandhi for launching a movement. But Gandhiji was not agreeable. This was the last meeting between Gandhiji and Subhas Bose. While in Europe and Far East, Subhas Bose watched the political developments in India closely and commented upon them from time-to-time. He praised the Quit India Movement sponsored by Mahatma Gandhi which enhanced his esteem a hundred fold.

REFERENCES

1. Bose, Subhas Chandra, *The Indian Struggle*, 1920-1942, *op. cit.*, p. 293.
2. *Ibid.*, p. 294.
3. *Ibid.*, pp. 294-95.
4. *Ibid.*, p. 295.
5. *Ibid.*, p. 70.
6. *Ibid.*, p. 74.
7. Kapoor, A.N. and V.R Gupta, *A Dictionary of Gandhian Thought* (Delhi: Ambe Books), 1995, pp. 211-12.
8. *Ibid.*, p. 179 80.
9. *Ibid.*, p. 182.
10. *Ibid.*, pp. 182-83.
11. *Ibid.*, p. 248.
12. Bose, Sisir K. and Sugata Bose (ed.), *The Essential Writings of Netaji Subhas Chandra Bose* (Delhi: Oxford University Press), 1997, p. 219.

13. The title of the essay is Forward Block: Its justification. It was one of the documents delivered personally to Sarat Chandra Bose in Calcutta by Bhagat Ram Talwar, Netaji's escort from Peshawar to Kabul. For the text of document see Subhas Chandra Bose, *op. cit.*, pp. 395-418.
14. Bose, Sisir K and Sugata Bose (ed.), *op. cit*, pp. 300-09.
15. Subhas Chandra Bose, *The Indian Struggle, op. cit.*, pp. 53-55.
16. *Ibid.*, pp. 61-62.
17. *Ibid.*, pp. 73-74.
18. *Ibid.*, pp. 69-74.
19. *Ibid.*, p. 70.
20. *Ibid.*, p. 175.
21. *Ibid.*, pp. 208-09.
22. *Ibid.*, pp. 209-31.
23. *Ibid.*, pp. 218-231.
24. *Ibid.*, pp. 248-49.

BIBLIOGRAPHY

Adhikari, G.D., (ed.), *Documents of the History of the Communist Party of India*, 3 Vols. (New Delhi: People's Publishing House), 1971-82.

Adiraju, V.R., *Netaji Subhas Bose The Great Revolutionary* (Hyderabad: Om See Satya Publications), 2001.

Adorno, Theodore (ed.), *The Authoritarian Personality* (New York, Harper and Row), 1950.

Aiyar, S.P. (ed.), *The Politics of Mass Violence in India* (Bombay: P.C. Manaktala & Sons), 1967.

Ambedkar, B.R., *Pakistan or The Partition of India* (Lahore: Mustafa Waheed), 1976.

Ambedkar, B.R., *What Congress and Gandhi Have Done to the* Untouchables (Bombay: Thacker & Co.), 1946.

Anand, Y.P., *The Essential Relationship Between Netaji Subhas Chandra Bose and Mahatma Gandhi* (New Delhi: National Gandhi Museum), 1997.

Argov, Daniel, *Moderates and Extremists in the Indian National Movement*, 1883-1923, Bombay, 1968.

Ayer, S.A., *Unto Him a Witness* (Bombay: Thacker & Co.), 1951.

Ayer, S.A., *Selected Speeches of Subhas Chandra Bose* (New Delhi: Publications Division, Ministry of Information and Broadcasting, Government of India), September 1974.

Azad, Maulana Abul Kalam, *India Wins Freedom* (Madras: Orient Longman), 1989.

Ba Maw, *Breakthrough in Burma, Memoirs of a Revolution, 1939-1946* (London: Yale University Press), 1968.

Bakshi, S.R., *V.D. Savarkar: Symbol of Sacrifice* (New Delhi: Anmol), 1994.

Basu, Aparna, *The Growth of Education and Political Development in India* 1898-1920 (Delhi: Oxford University Press), 1974.

Bell Charles, *Indian Mutiny* (London: Clarendon Press), 1931.

Bhattacharya, Buddhadeva (ed.), *Freedom Struggle and Anushilan Samiti* (Calcutta: Anushilan Samity), 1979.

Bimal Prasad (ed.), *A Revolutionary Quest* (Delhi: Oxford University Press), 1980.

Bose, Arun Coomar, *Indian Revolutionaries Abroad, 1905-1922* (Patna: Bharati Bhavan), 1971.

Bose, Asoke Nath, *My Uncle Netaji* (Calcutta: Esem), 1977.

Bose, Mihir, *The Lost Hero* (London: Quartet Books), 1982.

Bose, Nirmal Kumar, *Studies in Gandhism* (Calcutta: Merit), 1962.

Bose, Nirmal, *My Days with Gandhi* (Calcutta: A. Chatterjee), 1953.

Bose, Sarat Chandra, *I Warned My Countrymen* (Calcutta: Netaji Research Bureau), 1968.

Bose, Subhash Chandra, *An Indian Pilgrim: An Unfinished Autobiography and Collected Letters* (Bombay: Asia Publishing House), 1965.

Bose, Subhas Chandra, *Dissentient Report* (Kodalia: S.C. Bose), 1961.

Bose Subhas Chandra, *Impression in Life* (Lahore: Hero Publication), 1947.

Bose, Subhas Chandra, *Netaji Collected Works*, 12 Vols. (Calcutta: Netaji Research Bureau), 1980-1996.

Bose, Subhas Chandra, *The Indian Struggle* (Calcutta: Published for Netaji Publishing Society), 1948.

Bose, S.C., *Samagra Rachanabali*, Vol. 1 (Calcutta: Ananda Publishers), 1980.

Bose, Sugata, *Peasant Labour and Colonial Capital in Rural Bengal Since 1970* (Calcutta Ananda Publishers).

Bose, Sisir K., *Remembering My Father: A Centenary Offering* (Calcutta: Netaji Research Bureau), 1988.

Bose, Sisir K., *The Great Escape* (Calcutta: Netaji Research Bureau), 1975.

Bose, Sisir K., *Netaji and India's Freedom* (Calcutta: Netaji Bhavan), 1975.

Bose, Sisir K., and Alexander, Werth (eds.), *A Beacon Across Asia* (Delhi: Orient Longman), 1972.

Bose, Sisir K., and Sugata Bose (eds.), *The Essential Writings of Netaji Subhas Chandra Bose* (Calcutta: Oxford University Press), 1997.

Bose, S.K., *Tej Bahadur Sapru* (New Delhi: Publications Division, Ministry of Information & Broadcasting, Government of India), 1978.

Brass, Paul R., *Language, Religion and Politics in North India* (Delhi: Vikas), 1975.

Brown, Judith M. (ed.), *Eminent Indians* (Calcutta: Shanti Mitra), 1946.

Burham, James, *The Managerial Revolution* (England: Penguin Books Ltd.), 1942.

Chakarvarty S.R., and Madan C. Paul, *Netaji Subhas Chandra Bose: Relevance to Contemporary World* (New Delhi: Har-Anand), 2000.

Carr, E.H., *International Relations Between the Two World Wars* (London: MacMillan & Co. Ltd.), 1952.

Cashman, Richard, *The Myth of the Lokamanya Tilak and Mass Politics in Maharashtra* (Berkeley: University of California Press), 1975.

Chandra, Bipan, *Communalism in India* (New Delhi: Vani Educational Books), 1984.

Chandra, Bipan, India's Struggle for Independence 1857-1947 (New Delhi: Viking), 1998.

Chandra, Bipan, *Modern India* (New Delhi: NCERT), 1976.

Chandra, Bipan, *Nationalism and Colonialism in Modern India* (New Delhi: Orient Longman), 1979.

Chatterji, Dilip Kumar, *C.R. Das and Indian National Movement* (Calcutta: Post graduate Book Mart), 1965.

Chatterjee, Reva, *Netaji Subhas Bose, Bengal Revolution and Independence* (New Delhi: Ocean Books Pvt. Ltd.), 2000.

Chatterjee, Santimony (ed.), *Collected Works of Meghnath Saha* (Bengali), (Calcutta: Asia Publishing House), 1993.

Chattopadhyaya, Sarat Chandra, *Sharat Sahitya Samagra*, Vol. 2 (Calcutta: Anand Publisher), 1986.

Chaudhary, *Sukhbir, Growth of Nationalism in India*, 2 Vols. (New Delhi: Trimurti Publications), 1973.

Chirol, Valentine, *The Indian Unrest* (New Delhi: Lights and Life Publications), 1979.

Chirol, Valentine, *India Old and New* (New Delhi: Lights and Life Publications), 1975.

Chitkara, M.G., *Converts Do Not Make a Nation* (New Delhi: APH Publishing Company), 1998.

Collins, Larry and Lapierre, *Dominique, Freedom at Midnight* (New Delhi: Vikas Publishing House), 1997.

Copra, P.N. (ed.), *Quit India Movement: British Secret Report* (Faridabad: Thomson Press), 1976.

Dalton, Dennis, *Indian Idea of Freedom* (New Delhi: Manhohar), 1982.

Das, Cittaranjan, *India for Indians* (Madras: S. Ganesh), 1921.

Das, Durga, *India from Curzon to Nehru* (London: Collins), 1959.

Das, Harihar, *Subhas Chandra Bose and the Indian National Movement* (New Delhi Indian Book Co.), 1983.

Dasgupta, Hemendranath, *Deshbandhu Chittaranjan Das* (Delhi: Government of India, Publications Division), 1960.

Datta, K.K., *Renaissance, Nationalism and Social Changes in Modern India* (Calcutta: Bookland), 1965.

Datta, V.N., *Madan Lal Dhingra and the Revolutionary Movement* (New Delhi: Vikas Publishing House), 1978.

Dawarkadas, Kanji, *India's Fight for Freedom: An Eye Witness Story, 1913-37* (Bombay: Popular Prakashan), 1960.

Desai, A.R., *Social Background of Indian Nationalism* (Bombay: Popular Prakashan), 1976.

Dharamvira, *Lala Hardayal and Revolutionary Movements of His Times* (New Delhi: Indian Book Company), 1970.

Dignon, Don, *The Indian Revolutionary Problem in British Diplomacy 1914-1919* (New Delhi: Allied Publishers), 1983.

Dwivedi, Surendranath, *The Quest for Socialism* (New Delhi: Radiant), 1984.

Elsbree, Willard H., *Japan's Role in Southeast Asian Nationalist Movements, 1940 to 1945* (New York: Russel and Russel), 1953.

Fawcett, C.B., *Frontiers—A Study in Political Geography* (London: OUP), 1921.

Gandhi, M.K., *An Autobiography or The Story of My Experiments with Truth* (Ahmedabad: Navajivan Publishing House), 1983.

Gandhi, M.K., *Hind Swaraj* (Ahmedabad: Navajivan Publishing House), 1983.

Ganpule, N.G., *Netaji in Germany* (Bombay: Bhartiya Vidya Bhavan), 1959.

Ghosh, Chitra, "*The Socio-Political Ideas of Subhas Chandra Bose*" (Unpublished Paper, Presented in the Seminar of J.N.U. on 10-11 November, 1997).

Ghosh, Kalicharan, *The Roll of Honours: Ancedotes of Indian Martyrs* (Calcutta: Vidya Baharati), 1975.

Gooch, G.P., *Nationalism* (London: Longman), 1945.

Gopal, Madan, *Life and Times of Subhas Chandra Bose* (New Delhi: Vikas Publishing House Pvt. Ltd.), 1978.

Gordon, Leonard A., *Bengal: The Nationalist Movement, 1876-1940* (New York: Columbia University Press), 1974.

Gordon, Leonard, A., *Brothers Against the Raj* (Delhi: Viking), 1990.

Grover, Verinder, *Subhas Chandra Bose* (Delhi: Deep & Deep), 1999.

Guha, A.C. (ed.), *The Story of Indian Revolution* (Calcutta: Asia Publishing House), 1972.

Guha, A.C., *First Spark of Revolution* (Delhi: Orient Longman), 1972.

Guha, R. (ed.), *Subaltern Studies III* (Delhi: OUP), 1989.

Guha, Samar, *Netaji Dead or Alive* (New Delhi: S. Chand & Company), 1978.

Gupta, A.C. (ed.), *Studies in the Bengal Renaissance, in Commemoration of the Birth of Bipin Chandra Pal* (Calcutta: Asia Publishing House), 1958.

Gupta, Vishwa, Prakash, and Mohini Gupta, *Mahatma Gandhi: Vyakti aur Vichar* (Hindi), (New Delhi: Radha Publications), 1995.

Gupta, Vishwa Prakash and Mohini Gupta, *Subhas Chandra Bose: Vyakti aur Vichar* (Hindi) (New Delhi: Radha Publications), 1997.

Gupta, V.P. and Mohini Gupta, *The Quest for Freedom—A Study of Netaji Subhas Chandra Bose* (New Delhi: Radha Publications), 1998.

Gwyer, Sir Maurice and Appadorai, A., *Speeches and Documents on the Indian Constitution 1921-1947*, 2 Vols. (Delhi: Oxford University Press), 1957.

Haithcox, John P., *Communism and Nationalism in India: M.N. Roy and Comintern Policy, 1920-1939* (Princeton: Princeton University Press), 1971.

Haniz L. and Others, *The Individual Psychology of AlfredAdler* (New York, Harper and Row), 1967.

Harbick, Walter and Alexander Werth, *Netaji in Germany* (Calcutta: Netaji Research Bureau), 1970.

Hardayal, *Hints for Self Culture* (Delhi: Rajkamal Publications), 1948.

Hasan, Mushirul, *Nationalism and Communal Politics in India 1916-1928* (New Delhi: Manohar Publishers), 1981.

Hasan, Mushirul, *Communal Pan-Islamic Trends in Colonial India 1916-1928* (New Delhi: Manohar Publishers), 1981.

Hayashida, Tatsuo, *Netaji Subhas Chandra Bose* (Bombay: Allied), 1970.

Hitler, Adolph, *Mein Kampf* (Boston: Houghton Mifflin), 1943.

Holmes, T.R., *History of Indian Mutiny* (London: Macmillan), 1898.

Hugh, Toya, *Springing Tiger: A Study of Subhas Chandra Bose* (Calcutta: International Netaji Seminar), 1976.

Jinnah, M.A., *Speeches and Writings* (Lahore: Mustaffa), 1960.

Jog, N.C., *Lokamanya Bal Gangadhar Tilak* (New Delhi: Publications Division, Ministry of Information Broadcasting, Government of India), 1990.

Josh, Sohan Singh, *Hindustan Ghadar Party—A Short History* (New Delhi: People Publishing House), 1977.

Kapoor, A.N. and V.R Gupta, *A Dictionary of Gandhian Thought* (Delhi: Ambe Books), 1995.

Karan Singh, *Prophet of Indian Nationalism* (Bombay: Bharatiya Vidya Bhawan), 1991.

Karunakaran, K.P. (ed.), *Modern Indian Political Thought* (New Delhi; Allied Publishers), 1962.

Kaushik, P.D., *The Congress Ideology and Programme, 1920-47* (New Delhi: Allied), 1964.

Ker, James Compbell, *Political Trouble in India* (Calcutta: Superintendent Government Printing), 1917.

Khan, Shahnawaz, *My Memories of INA and its Netaji* (Delhi: Rajkamal Publication), 1946.

Khan, Shah Nawaz, *I.N.A. and Its Netaji* (Delhi: Allied Publishers), 1946.

Laski, Harold, *Grammar of Politics* (London: Allen), 1948.

Lebra, Joyce C., *Jungle Alliance: Japan and the Indian National Army* (Singapore: Asia Pacific Press), 1959.

Lohia, Rammanohar, *Guilty Men of India's Partition* (Allahabad: Kitabistan), 1960.

Maikap, S.C., *Challenge to the Empire: A Study of Netaji* (New Delhi: Publications Division, Government of India), 1995.

Majumdar, R.C., *History of the Freedom Movement in India* (Calcutta: Firma K.L. Mukhopadhyaya), 1963.

Malleson, G.B. and Kye John, *History of Indian Mutiny* (London: Allen), 1879.

Masani, R.P., *Dadabhai Naoroji* (New Delhi: Publications Division, Ministry of Information Broadcasting, Government of India), 1960.

Masselos, Jim, *Indian Nationalism: An History* (New Delhi: Sterling), 1985.

Mathur, L.P., *Indian Revolutionary Movement in USA* (Delhi: S. Chand), 1970.

Mazumdar, R.C., Roy Chaudhary, H.C. Dutta, Kali Kumar, *An Advanced History of India* (Madras: Macmillan), 1950.

Mazumdar, S.N., *In Search of a Revolutionary Ideology* (New Delhi: People Publishing House), 1979.

McCarthy, Justin, *History of Our Times*, Vol. II (London: Chatts), 1897.

Mead, Henery, *The Sepoy Revolt: Its Causes & Consequences* (Delhi: Gyan), 1986.

Menon, V.P., The *Transfer of Power in India* (Bombay: Orient Longman), 1957

Mody, Homi, *Pherozeshah Mehta* (New Delhi: Publications Division, Ministry of Information & Broadcasting, Government of India), 1963.

Mukherjee, Radhakamal, *India: The Dawn of New Era* (New Delhi: Radha Publications), 1997.

Nanda, B.R., Gandhi, *Pan-Islamism, Imperialism and Nationalism in India* (Bombay: Oxford University Press), 1989.

Nath, Shaileshwar, *Terrorism in India* (New Delhi: National Publishing House), 1980.

Nehru, Jawaharlal, *An Autobiography* (London: The Bodley Head), 1958.

Pandit, V.L., *The Scope of Happiness* (New Delhi: Vikas Publishing House), 1965.

Parmanand, Bhai, *The Story of My Life* (New Delhi: S. Chand), 1982.

Parekh, Bhikhu, *Colonialism, Tradition and Reform* (Delhi: Sage Publications), 1999.

Philips, C.H. (ed.), *The Evolution of India and Pakistan 1858-1947 Select Documents* (London: George Allen & Unwin), 1962.

Planning Commission, *Subhas Chandra Bose: Pioneer of Indian Planning* (New Delhi: Planning Commission, Government of India), 1997.

Radhakrishanan, S. (ed.), *Mahatma Gandhi* (London: George Allen), 1949.

Rai, Lala Lajpat, *The Story of My Life* (New Delhi: D.A.V. Publications), 1987.

Rai, Lala Lajpat, *Young India: An Interpretation and a History of the Nationalist Movement from Within* (New Delhi: Publications Division, Ministry of Information Broadcasting, Government of India), 1968.

Rajendra Prasad, *An Autobiography* (Bombay: Asia Publishing House), 1957.

Ralhan, O.P. (ed.), *Hindu Mahasabha*, 2 Vols. (New Delhi: Anmol), 1997.

Ray, Sibnarayan (ed.), *M.N. Roy, Philosopher and Revolutionary* (Delhi: Oxford University Press), 1987.

Reddy, P.L. Sanjeev and Mishra, S.N. (eds.), *Sardar Vallabhbhai Patel: The Nation Builder* (New Delhi: Indian Institute of Public Administration), 2004.

Rolland Romain, *The Life of Vivekananda and the Universal Gospel* (Calcutta: Advaita Ashrama), 1953.

Rolland, Romain, *Mahatma Gandhi,* translation from French by Catherine D. Groth (New Delhi: Publications Division, Government of India).

Romain Rolland and Gandhi Correspondence (New Delhi: Publications Division, Government of India), 1976.

Roy, Dilip Kumar, *The Subhash I Knew* (Bombay: Asia Publishing House), 1946.

Saggi, P.D. (ed.), *Life and Works of Netaji Subhas Chandra Bose* (Bombay: Overseas Publishing House), n.d.

Sarkar, M.L. (ed.), *Subhas Bose: 1939-40* (Bengali), Translate by Nipen Chakravarty (Calcutta: Forward Publishing Concern), 1975.

Sarkar, S., *Modern India: 1885-1947* (Delhi: Macmillan), 1983.

Savarkar, V.D., *Six Glorious Epics in Indian History* (New Delhi: Rajdhani Granthagar), 1970.

Savarkar, V.D., *Swantantryaveer Savarkar: A Guide to Indian Revolutionary Movement* (Pune: Maharashtra Prantik Hindu Sabha), 1963.

Savarkar, V.D., *The Indian War of Independence 1847* (New Delhi: Phoenix), 1947.

Savarkar, V.D., *Then Story of My Transportation of Life* (Bombay: Sadbhakti Publications), 1950.

Savarkar, V.D., *Echo from Andamans* (Poona: Venus Books), 1947.

Savarkar, S.S. and Joshi G.M., *Historic Statements V.D. Savarkar*, V.D (Bombay: Popular Prakashan), 1967.

Seale, Patrick and Maureen McConville, P. (London: Penguin Books Ltd.), 1978.

Sen, Rajen (ed.), *Deshbandhu Chitta Ranjan* (Calcutta: Ananda Publishers), 1927.

Shankar, Ramahari, *Gandhi's Encounter with Indian Revolutionaries* (New Delhi: Siddharth Publications), 1996.

Shankhdher, M.M., *Reflections on Indian Politics* (New Delhi: Kumar Brothers), 1973.

Sharma, Jai Narain, *Power, Politics and Corruption: A Gandhian Solution* (New Delhi: Deep and Deep), 2004.

Sharma, Malikarjuna, *Role of Revolutionaries in the Freedom Struggle* (Hyderabad: Marxist Study Forum), 1987.

Shraddhanand, Swami, *Inside Congress* (Bombay: Phoenix), 1946.

Singh, Gulab, *Under The Shadow of Gallows, A Story of a Revolutionary* (Delhi: Rup Chand), 1964.

Singhal, Jwala Prasad, *Sphinx Speaks: The Story of Pre-historic Nations* (New Delhi: Sadgya), 1963.

Sitaramayya, Pattabhi, *History of Indian National Congress* (Delhi: S. Chand & Co.), 1969.

Sri Aurobindo, *Sri Aurobindo on Himself* (Pondicherry: Sri Aurobindo Ashram), 1972.

Srivastava, Harindra, *Five Stormy Years: Savarkar in London* (New Delhi: Allied Publishers), 1983.

Suda, J.P., *Main Currents of Social and Political Thought in India*, Vol. III (Meerut: K. Nath & Co.), 1973.

Sullivan, O. Noel (ed.), *Terrorism, Ideology and Revolution* (Sussex: The Harvester Press), 1986.

Swami Jyotiramayananda (ed.), *Vivekananda, With a Garland of Tributes* (Madras: S. Ganeshan), 1988.

Swami Nikhilananda, *Vivekananda* (Calcutta; Advaita Ashram), 1999.

Tara Chand, *History of Freedom Movement in India* (New Delhi: Publications Division, Ministry of Information & Broadcasting, Government of India), 1962.

Tendulkar, D.G., *Mahatma*, 8 Vols. (New Delhi: Publications Division, Ministry of Information & Broadcasting, Government of India), 1958.

The Collected Works of Mahatma Gandhi, 100 Vols. (New Delhi: Government of India, Publications Division), 1958-1995.

Toye, Hugh, *Subhash Chandra Bose: The Springing Tiger* (Delhi: Jaico Publishing House), 2001.

Trehan, Jyoti, *Veer Savarkar* (New Delhi: Deep and Deep), 1991.

Varma, V.P., *Modern Indian Political Thought* (Agra: Luxmi Narain Agarwal), 1961.

Vajpeyi, J.N., *The Extremist Movement in India* (Allahabad: Chug Publications), 1974.

Wardlow, Grant, *Political Terrorism Theory, Tactics and Counter Measures* (Cambridge: Cambridge University Press), 1982.

Walport, Stanley, *Jinnah of Pakistan* (New Delhi: Oxford University Press), 1965.

Walport, Stanley, *Tilak and Gokhale* (Bombay: University of California Press), 1962.

Wilkinson, Paul, *Political Terrorism* (New York: Macmillan), 1976.

Wasti, Syed Razi, *Lord Minto and the Indian National Movement 1905 - 1910* (London: Oxford University Press), 1965.

Yadav, K.C. and Akiko Sen (eds.), *Subhas Chandra Bose* (New Delhi: Hope India), 2003.

Zangwill, Israel, *Principles of Nationalities* (London: Heinemann), 1920.

Index

Abid Hussain 85
Action against British Advocacy 51
Action against General Dyer 21
Afghanistan 62
Ahimsa 105
Albert School 12
Alipur Central Jail 64
All India National Manning Committee 73
All India Naujawan Bharat Sabha 65, 67
All India Political Conference 42
All India Trade Union Congress 65
Allied Power 63
Ambedkar, B.R. 118
Andrew, C.F. 88
Anglo-American imperialism 55
Anilbaran Roy 26
Anti-imperialist Power 55
Anti-Fascist approach 53
Aristocracy 68
Armed revolution 105
Ata Turk Model 63
Aurobindo Ghose 17, 26, 28, 29
Axis power 30-32, 44, 51-53, 55-67
Azad Hind Fouz 55, 85
Azad Hind Radio 58

Ba Maw (Prime Minister of Burma) 32
Bahadur Shah 84
Banerjee, Satkar 12
Banerjee, Surendranath 17
Banglar Katha 41
Baptist Mission 13
Bardoli 47
Battle of Plassey 83
Bengal Criminal Law Amendment Act 33
Bengal Pact 79, 80
Bengal Pradesh Committee 53
Bengal Provincial Congress Committee 80
Bengal Youth Conference 35, 41
Benito Mussolini 63-64
Bhadralok 37
Bhagat Singh 115
Bharatiya Naujawan Sabha 67
Bhowali's sanatorium 26
Bolshevism 64, 68, 90
Bose, Subhas Chandra,
  conceptual ideas 58-59
  family environment 11-14
  philosophy and ideas 46-47
  thoughts and ideas 40-41
  views on fascism 88-100
  views on nation building 47
Boycott and passive resistance 50
Brahmacharya 15, 105
Brahman 6, 15
Brahmo Samaj 12, 13
British administration 45
British Commonwealth 57
British imperialism 43, 62
British Raj 42
British Rule 13, 21
Burke, Edmund 105

Calcutta Corporation election 37
Caste system 48, 76
Centralization 74-75
Chalo Delhi 44

Chattopadhyaya, Khitish Prasad 41
Chattopadhyaya, Sarat Chandra 38, 67
Chattopadhyaya, Virendranath 93
Chauri Chaura 111
China 62
Chittagong 64
Chittaranjan Das 29, 33, 37, 40
Civil Disobedience movement 46, 90, 109, 116
Conflict between cottage industries and Large Scale industries 74
Congress-League alliance 112
Communal amity 79-86
  principle of 82-84
Communal Award 118
Communal riots 112
Communalism 48-50
Communism 69
Co-Prosperity Sphere 57
Congress Working Committee 42, 71, 120
Cripps scheme 30
Cultural and Commercial Association 36
Currency and exchange 34

Daridranarayan 64
Das, C.R. 22, 26, 29-30, 33-36, 38, 64-65, 79-82, 93, 106
Deoli camp 96
Delhi Pact 47
Dharma 25
Dharmayuddha 43
Dictatorship 68, 98, 100
Dutt, Rajani Palme 99
Dwarkanath Vidyasagar 12

Economic Consciousness 82
Economic improvement 42
Economic programme 83
Economic self-sufficiency 35
England 117

Fascism 50, 65, 68
Fascist Party 100
First World War 63, 97
Five Year Plan system 42
Foreign policy 52
Forward Block 98, 109
Freedom intoxicated missionaries 47
Freedom, concept of 27, 34, 48-49
Full-blooded socialism 65-66

Gandhi-Irwin Pact 90, 109, 114, 115, 116
Gandhian Leadership, Critique of 104-20
  negative aspects of 110-11
Gandhian Mass politics 38
Gandhian movement 109-10
German Nazis 89-90
Gita 15, 24
Guerilla Warfare 58

Hachiya (Japanese Ambassador) 57
Hindu Mahasabha 37, 49,86
Hindu Muslim Unity 79-80, 83-86, 112
Hitler 56, 89, 93, 99
Ho chi Minh 44
Hooghly District Student's Conference 70
House of Commons 21

Ideology 24-30
Imphal operation 55, 56
India, freedom struggle of 30, 104
  liberation movement 43, 54
Indian National Army 32, 45, 84
Indian National Congress (INC) 21, 43, 52, 98, 100, 105-06, 109-10
Indian Nationalism 19-21
Indian Revolution 21-22
Industrialization 42, 71-74
Italian Fascism 63-65
Italy 104
Iyenger, Rangaswami 106

Janakinath Bose (Subhas' father) 11-12
Japan 57, 62
Jessore Provincial Conference 35
Jinnah, Mohammad Ali 49, 86, 107

Karachi's Naujawan Bharat Sabha
  Conference 65
Karl Marx 65
Kelkar, N.C. 106
Khilafat issue 111
Krishna Kriplani 89
Kshatriya 15

Lahore 48
Lahore Congress 109
Lala Lajpat Rai 33, 106
Leftism 77
Leviathan 16
Liberation 42
Lokmanya Tilak 106
Lord Irwin 114

Mahabharata 105
Maharashtra Political Conference 65, 82
Mahatma Gandhi 33-35, 81-85,
  91-93, 99-102, 105-10, 115-20
  march to Dandi 108
  visit to England 117
Mahmud uz Zafar 96
Man of Destiny 58
Manchuria 62
Mandalay Jail 24, 30, 33, 37, 81
Mantra-Shakti 27
Mao Zedong 44
Marxian socialism 69
Marxism-Leninism 65
Maulana Shaukat Ali 107
Maulana Abul Kalam Azad 107
McDonald, Ramsay 118
Militant Nationalism 46-47
Mission School 14, 101
Modern Scientific development 35, 36
Moscow Communists 90, 93
Mukherji, Ashutosh 7
Munich Pact 119
Muslim aristocracy 84
Muslim League 49, 86
Mussolini 88-90, 92, 97-100, 108
Mustafa Kamal Pasha 63, 111

Napoleon 46, 108
Nation Building programmes 69
National liberation movement 63
National Planning Commission 41, 71
Nationalism 19, 43-45, 47-48, 75-76
Nawab Siraj-ud-Daulah 84
Nazi-Fascist gangs 94
Nazism 56
Nehru, Jawaharlal 64-68, 72-74,
  93-96, 98-101, 106-07, 113-15
Nehru, Motilal 34, 106, 112
Nishkama Karma 28
Non-Cooperation movement 26, 97,
  107, 110-12, 114
Non-Violent guerrilla warfare 50

Order of the Rising Sun 57
Orthodox Brilliance, sacrifice of 212

Palli-samaj 37
Panchayat system 76
Pandit, Vijaya Lakshmi 89, 96
Pant, Govind Ballabh 106
Partition of India 50
Plutocracies 63
Poona agreement 106, 114
Poona Pact 118
Pondicherry ashram 26
Poverty 41, 48, 71, 73
Presidency College 17
Propaganda 36, 40
Purandarpur 12

Quit India movement 50, 110, 120

Rajendra Prasad 94, 107
Rajagopalachari 106
Ramakrishna 16, 17, 32
Ramakrishna Mission 32
Ramakrishna Paramhamsa 13, 15
Ramayana 105
Ramrajya 105
Rangpur political conference 65, 82
Rani of Jhansi 49
Ravenshaw Collegiate School 14
Real politik 52
Regional autonomy 74-75

Regulation, II, III 53, 64
Revolutionary movement 17, 33
Rolland, Romain 88-90, 92-96, 98
Roman Catholic Bishop 91
Round Table Conference 90, 114-16, 118
Roy, Dilipkumar 20, 24
Roy, M.N. 12
Roy, P.C. 17
Roy, Raja Ram Mohun 13, 32
Rumour-mongering 36
Russia 56, 98, 104
Russo-German War 54

Sadhana 25
Sadhu hunting 28
Saheed (martyr) 57
Samyavada 85, 97
Samyavada Sangh 50
Sanatana Dharma 31
Sardar Ajit Singh 24
Satyagrahi 101
Scottish Church College 12
Second Round Table Conference 26
Second World War 30, 43, 54, 66, 97, 120
Secularism 50
Sen, Keshav Chandra 13
Shah, Meghnath 41, 72
Shahnawaz Khan 84
Singapore 85
Sirajuddowla Day 49
Small Scale Village and Khadi Industries 73
Socialism 64-76
Socialist Republic of India 67
Soviet Russia 72, 90
Spencer, Herbert 105
Spiritualism 27, 35
St. Francis of Assissi 89
St. Xavier's College 12
Struggle against British Rule 43, 50
Student's Consultative Committee 19
Sudras 15
Sultan Hussain Shah 12
Swadeshi 33, 47-48
Swadharma 25, 31
Swami Vivekananda 15-17, 25, 30-33, 34, 65, 93
Swaraj 21, 30, 34, 57, 99
Swaraj Party 37, 41, 96

Tagore, Rabindranath 13, 17, 34, 89, 92
Tajo (Japan's Prime Minister) 32
Teacher's Training Institute 41
Terrorism 45
Thoreau 106
Tito 44
Tokyo 54
Tolstoy, Leo 106
Tsarist Russia 63
Turkey 62, 63

United National Party 117
University of Tokyo 76
Upanishad 15
USSR 52

Vaisyas 15
Vallabhbhai Patel 106
Versailles Peace Treaty 63
Vidyasagar, Ishwar Chandra 13
Vienna 26
Vietnam war 44
Violent revolution 19
Vitthal Bhai Patel 106
Voluntary Corps 14
Vo Nguyen Giap 44

Wafd Party 62
Wardha resolution 70

Yoga 15
Yogi 27
Youth organisation 19

Zamindari system 76